# Materials, Processes, and Reliability for Advanced Interconnects for Micro- and Nanoelectronics — 2009

MATERIALS RESEARCH SOCIETY
SYMPOSIUM PROCEEDINGS VOLUME 1156

# Materials, Processes, and Reliability for Advanced Interconnects for Micro- and Nanoelectronics — 2009

Symposium held April 14–17, 2009, San Francisco, California, U.S.A.

### EDITORS:

## Martin Gall
Freescale Semiconductor
Hopewell Junction, New York, U.S.A.

## Alfred Grill
IBM T.J. Watson Research Center
Yorktown Heights, New York, U.S.A.

## Francesca Iacopi
IMEC
Leuven, Belgium

## Junichi Koike
Tohoku University
Sendai, Japan

## Takamasa Usui
Toshiba America Electronic Components, Inc.
Albany, New York, U.S.A.

Materials Research Society
Warrendale, Pennsylvania

CAMBRIDGE UNIVERSITY PRESS
Cambridge, New York, Melbourne, Madrid, Cape Town,
Singapore, São Paulo, Delhi, Mexico City

Cambridge University Press
32 Avenue of the Americas, New York NY 10013-2473, USA

Published in the United States of America by Cambridge University Press, New York

www.cambridge.org
Information on this title: www.cambridge.org/9781107408319

Materials Research Society
506 Keystone Drive, Warrendale, PA 15086
http://www.mrs.org

First published 2009
First paperback edition 2012

Single article reprints from this publication are available through
University Microfilms Inc., 300 North Zeeb Road, Ann Arbor, MI 48106

CODEN: MRSPDH

ISBN 978-1-107-40831-9 Paperback

Preface ..................................................................................................................ix

Materials Research Society Symposium Proceedings.............................................x

## LOW-k DIELECTRICS I

Effect of Trapping on Dielectric Conduction Mechanisms
of ULK/Cu Interconnects..........................................................................................3
    Virginie Verriere, Cyril Guedj, David Roy,
    Serge Blonkowski, and Alain Sylvestre

Dual Damascene Reactive Ion Etch Polymer Characterization
Through X-ray Photoelectron Spectroscopy for 65 nm and
45 nm Technology Nodes.........................................................................................11
    Samuel Choi, Leo Tai, and Chet Dziobkowski

Interaction of O and H Atoms with Low-k SiOCH Films
Pretreated in He Plasma..........................................................................................17
    O.V. Braginsky, A.S. Kovalev, D.V. Lopaev,
    Y.A. Mankelevich, T.V. Rakhimova, E.M. Malykhin,
    O.V. Proshina, A.T. Rakhimov, A.N. Vasilieva,
    D.G. Voloshin, S.M. Zyryanov, and Mikhail R. Baklanov

Characterization of Plasma Damage in Low-k Films by
TVS Measurements ..................................................................................................23
    Ivan Ciofi, Mikhail R. Baklanov, Giovanni Calbo,
    Zsolt Tökei, and Gerald Beyer

## LOW-k DIELECTRICS II

Effects of Polymer Material Variations on High Frequency
Dielectric Properties ...............................................................................................31
    Gregory Pawlikowski

Optimization of Low-k UV Curing: Effect of Wavelength on
Critical Properties of the Dielectric.......................................................................39
    German Aksenov, Patrick Verdonck, David DeRoest,
    F.N. Dultsev, Premysl Marsik, Denis Shamiryan,
    H. Arai, N. Takamure, and Mikhail R. Baklanov

Application of UV Irradiation in Removal of Post-Etch
193 nm Photoresist..............................................................................................45
    Quoc Toan Le, Els Kesters, L. Prager, Marcel Lux,
    P. Marsik, and Guy Vereecke

*POSTER SESSION: INTERCONNECTS*

Effects of Silica Sources on Nanoporous Organosilicate
Films Templated with Tetraalkylammonium Cations..........................................53
    Salvador Eslava, Jone Urrutia, Abheesh N. Busawon,
    Mikhail R. Baklanov, Francesca Iacopi, Karen Maex,
    Christine E. Kirschhock, and Johan A. Martens

Electrical and Structural Properties of Ultrathin Polycrystalline
and Epitaxial TiN Films Grown by Reactive dc Magnetron
Sputtering ............................................................................................................59
    Fridrik Magnus, Arni S. Ingason, Sveinn Olafsson, and
    Jon T. Gudmundsson

A Study of Diffusion Barrier Characteristics of Electroless
Co(W,P) Layers to Lead-Free SnAgCu Solder .................................................65
    Hung-Chun Pan and Tsung-Eong Hsieh

*METALLIZATION I*

Atomic Layer Deposition of Ruthenium Films on Hydrogen
Terminated Silicon..............................................................................................73
    Sun Kyung Park, K. Roodenko, Yves J. Chabal,
    L. Wielunski, R. Kanjolia, J. Anthis, R. Odedra,
    and N. Boag

Coupled Finite Element — Potts Model Simulations of
Grain Growth in Copper Interconnects................................................................79
    Bala Radhakrishnan and Gorti Sarma

Electroless Cu Deposition on Self-Assembled Monolayer
Alternative Barriers.............................................................................................85
    Silvia Armini and Arantxa Maestre Caro

Rutherford Backscattering Spectrometry Analysis of
Growth Rate and Activation Energy for Self-Formed
Ti-Rich Interface Layers in Cu(Ti)/Low-k Samples ...................................... 93
    Kazuyuki Kohama, Kazuhiro Ito, Kenichi Mori,
    Kazuyoshi Maekawa, Yasuharu Shirai, and
    Masanori Murakami

Adhesion and Cu Diffusion Barrier Properties of a MnO$_x$
Barrier Layer Formed with Thermal MOCVD ....................................... 99
    Koji Neishi, Vijay K. Dixit, S. Aki, Junichi Koike,
    K. Matsumoto, H. Sato, H. Itoh, and S. Hosaka

Electronic Transport Properties of Cu/MnO$_x$/SiO$_2$/p-Si
MOS Devices ........................................................................................... 105
    Vijay K. Dixit, Koji Neishi, and Junichi Koike

## METALLIZATION II

Stress Gradients Observed in Cu Thin Films Induced by
Capping Layers ....................................................................................... 113
    Conal E. Murray, Paul R. Besser, Christian Witt,
    and Jean L. Jordan-Sweet

## RELIABILITY

* Large-Scale Electromigration Statistics for Cu Interconnects .......................... 121
    Meike Hauschildt, Martin Gall, and Richard Hernandez

Effect of Dielectric Capping Layer on TDDB Lifetime of
Copper Interconnects in SiOF .......................................................... 133
    Jeff Gambino, Fen Chen, Steve Mongeon, Phil Pokrinchak,
    John He, Tom C. Lee, Mike Shinosky, and Dave Mosher

## EMERGING INTERCONNECT
## TECHNOLOGIES

Direct Metal Nano-Patterning Using Embossed Solid Electrolyte ................... 141
    Anil Kumar, Keng Hsu, Kyle Jacobs, Placid Ferreira,
    and Nicholas Fang

*Invited Paper

## JOINT SESSION:
## INTERCONNECT AND PACKAGING

**Low Temperature Direct Cu-Cu Immersion Bonding for 3D Integration** .................................................................................149
  Rahul Agarwal and Wouter Ruythooren

**Failure Analysis and Process Improvement for Through Silicon Via Interconnects** ...............................................................155
  Bivragh Majeed, Marc Van Cauwenberghe,
  Deniz S. Tezcan, and Philippe Soussan

**Effects of Thinned Multi-Stacked Wafer Thickness on Stress Distribution in the Wafer-on-a-Wafer (WOW) Structure** .............................................................................................163
  Hideki Kitada, Nobuyuki Maeda, Koji Fujimoto,
  Tomoji Nakamura, Kousuke Suzuki, and Takayuki Ohba

**\* Power Delivery, Signaling and Cooling for 3D Integrated Systems** ....................................................................................169
  Muhannad Bakir and Gang Huang

**Copper Deposition Technology for Thru Silicon Via Formation Using Supercritical Carbon Dioxide Fluids Using a Flow-Type Reaction System** ..............................................181
  Masahiro Matsubara and Eiichi Kondoh

**Author Index** ........................................................................................187

**Subject Index** ......................................................................................189

\*Invited Paper

# PREFACE

The semiconductor industry continues to follow Moore's law into 32 nm and 22 nm technologies, enabled by the development and introduction of new materials. Advanced interconnect structures require the use of porous dielectrics with further reduced k-values and even weaker mechanical properties, as well as much thinner metallization liners. In addition, the increasing resistivity of Cu with decreasing dimensions needs to be addressed in order to maintain the performance of the continuously shrinking devices. Innovations in materials, processes and architectures are needed to address these issues and to maintain the reliability of the interconnects.

Symposium D, "Materials, Processes, and Reliability for Advanced Interconnects for Micro- and Nanoelectronics —2009," held April 14–17 at the 2009 MRS Spring Meeting in San Francisco, California, aimed to provide a forum for researchers around the world to exchange the latest advances in materials, processes, integration and reliability in advanced interconnects and packaging, and to discuss interconnects for emerging technologies, including 3D chip stacking and molecular electronics. A joint session was held with Symposium F, "Packaging, Chip-Package Interactions and Solder Materials Challenges."

The editors gratefully acknowledge the support of Air Products and Chemicals Inc., Applied Materials Inc., IBM T.J. Watson Research Center, JSR Micro Inc., Novellus Systems Inc., and Tokyo Electron America Inc.

Martin Gall
Alfred Grill
Francesca Iacopi
Junichi Koike
Takamasa Usui

August 2009

# MATERIALS RESEARCH SOCIETY SYMPOSIUM PROCEEDINGS

Volume 1153 — Amorphous and Polycrystalline Thin-Film Silicon Science and Technology — 2009,
A. Flewitt, Q. Wang, J. Hou, S. Uchikoga, A. Nathan, 2009, ISBN 978-1-60511-126-1

Volume 1154 — Concepts in Molecular and Organic Electronics, N. Koch, E. Zojer, S.-W. Hla, X. Zhu,
2009, ISBN 978-1-60511-127-8

Volume 1155 — CMOS Gate-Stack Scaling — Materials, Interfaces and Reliability Implications,
J. Butterbaugh, A. Demkov, R. Harris, W. Rachmady, B. Taylor, 2009,
ISBN 978-1-60511-128-5

Volume 1156— Materials, Processes and Reliability for Advanced Interconnects for Micro- and
Nanoelectronics — 2009, M. Gall, A. Grill, F. Iacopi, J. Koike, T. Usui, 2009,
ISBN 978-1-60511-129-2

Volume 1157 — Science and Technology of Chemical Mechanical Planarization (CMP), A. Kumar,
C.F. Higgs III, C.S. Korach, S. Balakumar, 2009, ISBN 978-1-60511-130-8

Volume 1158E —Packaging, Chip-Package Interactions and Solder Materials Challenges, P.A. Kohl,
P.S. Ho, P. Thompson, R. Aschenbrenner, 2009, ISBN 978-1-60511-131-5

Volume 1159E —High-Throughput Synthesis and Measurement Methods for Rapid Optimization and
Discovery of Advanced Materials, M.L. Green, I. Takeuchi, T. Chiang, J. Paul, 2009,
ISBN 978-1-60511-132-2

Volume 1160 — Materials and Physics for Nonvolatile Memories, Y. Fujisaki, R. Waser, T. Li,
C. Bonafos, 2009, ISBN 978-1-60511-133-9

Volume 1161E —Engineered Multiferroics — Magnetoelectric Interactions, Sensors and Devices,
G. Srinivasan, M.I. Bichurin, S. Priya, N.X. Sun, 2009, ISBN 978-1-60511-134-6

Volume 1162E —High-Temperature Photonic Structures, V. Shklover, S.-Y. Lin, R. Biswas, E. Johnson,
2009, ISBN 978-1-60511-135-3

Volume 1163E —Materials Research for Terahertz Technology Development, C.E. Stutz, D. Ritchie,
P. Schunemann, J. Deibel, 2009, ISBN 978-1-60511-136-0

Volume 1164 — Nuclear Radiation Detection Materials — 2009, D.L. Perry, A. Burger, L. Franks,
K. Yasuda, M. Fiederle, 2009, ISBN 978-1-60511-137-7

Volume 1165 — Thin-Film Compound Semiconductor Photovoltaics — 2009, A. Yamada, C. Heske,
M. Contreras, M. Igalson, S.J.C. Irvine, 2009, ISBN 978-1-60511-138-4

Volume 1166 — Materials and Devices for Thermal-to-Electric Energy Conversion, J. Yang, G.S. Nolas,
K. Koumoto, Y. Grin, 2009, ISBN 978-1-60511-139-1

Volume 1167 — Compound Semiconductors for Energy Applications and Environmental Sustainability,
F. Shahedipour-Sandvik, E.F. Schubert, L.D. Bell, V. Tilak, A.W. Bett, 2009,
ISBN 978-1-60511-140-7

Volume 1168E —Three-Dimensional Architectures for Energy Generation and Storage, B. Dunn, G. Li,
J.W. Long, E. Yablonovitch, 2009, ISBN 978-1-60511-141-4

Volume 1169E —Materials Science of Water Purification, Y. Cohen, 2009, ISBN 978-1-60511-142-1

Volume 1170E —Materials for Renewable Energy at the Society and Technology Nexus, R.T. Collins,
2009, ISBN 978-1-60511-143-8

Volume 1171E —Materials in Photocatalysis and Photoelectrochemistry for Environmental Applications
and $H_2$ Generation, A. Braun, P.A. Alivisatos, E. Figgemeier, J.A. Turner, J. Ye,
E.A. Chandler, 2009, ISBN 978-1-60511-144-5

Volume 1172E —Nanoscale Heat Transport — From Fundamentals to Devices, R. Venkatasubramanian,
2009, ISBN 978-1-60511-145-2

Volume 1173E —Electofluidic Materials and Applications — Micro/Biofluidics, Electowetting and
Electrospinning, A. Steckl, Y. Nemirovsky, A. Singh, W.-C. Tian, 2009,
ISBN 978-1-60511-146-9

Volume 1174 — Functional Metal-Oxide Nanostructures, J. Wu, W. Han, A. Janotti, H.-C. Kim, 2009,
ISBN 978-1-60511-147-6

# MATERIALS RESEARCH SOCIETY SYMPOSIUM PROCEEDINGS

Volume 1175E —Novel Functional Properties at Oxide-Oxide Interfaces, G. Rijnders, R. Pentcheva,
J. Chakhalian, I. Bozovic, 2009, ISBN 978-1-60511-148-3

Volume 1176E —Nanocrystalline Materials as Precursors for Complex Multifunctional Structures through
Chemical Transformations and Self Assembly, Y. Yin, Y. Sun, D. Talapin, H. Yang, 2009,
ISBN 978-1-60511-149-0

Volume 1177E —Computational Nanoscience — How to Exploit Synergy between Predictive Simulations
and Experiment, G. Galli, D. Johnson, M. Hybertsen, S. Shankar, 2009,
ISBN 978-1-60511-150-6

Volume 1178E —Semiconductor Nanowires — Growth, Size-Dependent Properties and Applications,
A. Javey, 2009, ISBN 978-1-60511-151-3

Volume 1179E —Material Systems and Processes for Three-Dimensional Micro- and Nanoscale Fabrication
and Lithography, S.M. Kuebler, V.T. Milam, 2009, ISBN 978-1-60511-152-0

Volume 1180E —Nanoscale Functionalization and New Discoveries in Modern Superconductivity,
R. Feenstra, D.C. Larbalestier, B. Maiorov, M. Putti, Y.-Y. Xie, 2009,
ISBN 978-1-60511-153-7

Volume 1181 — Ion Beams and Nano-Engineering, D. Ila, P.K. Chu, N. Kishimoto, J.K.N. Lindner,
J. Baglin, 2009, ISBN 978-1-60511-154-4

Volume 1182 — Materials for Nanophotonics — Plasmonics, Metamaterials and Light Localization,
M. Brongersma, L. Dal Negro, J.M. Fukumoto, L. Novotny, 2009,
ISBN 978-1-60511-155-1

Volume 1183 — Novel Materials and Devices for Spintronics, O.G. Heinonen, S. Sanvito, V.A. Dediu,
N. Rizzo, 2009, ISBN 978-1-60511-156-8

Volume 1184 — Electron Crystallography for Materials Research and Quantitative Characterization of
Nanostructured Materials, P. Moeck, S. Hovmöller, S. Nicolopoulos, S. Rouvimov,
V. Petkov, M. Gateshki, P. Fraundorf, 2009, ISBN 978-1-60511-157-5

Volume 1185 — Probing Mechanics at Nanoscale Dimensions, N. Tamura, A. Minor, C. Murray,
L. Friedman, 2009, ISBN 978-1-60511-158-2

Volume 1186E —Nanoscale Electromechanics and Piezoresponse Force Micropcy of Inorganic,
Macromolecular and Biological Systems, S.V. Kalinin, A.N. Morozovska, N. Valanoor,
W. Brownell, 2009, ISBN 978-1-60511-159-9

Volume 1187 — Structure-Property Relationships in Biomineralized and Biomimetic Composites,
D. Kisailus, L. Estroff, W. Landis, P. Zavattieri, H.S. Gupta, 2009,
ISBN 978-1-60511-160-5

Volume 1188 — Architectured Multifunctional Materials, Y. Brechet, J.D. Embury, P.R. Onck, 2009,
ISBN 978-1-60511-161-2

Volume 1189E —Synthesis of Bioinspired Hierarchical Soft and Hybrid Materials, S. Yang, F. Meldrum,
N. Kotov, C. Li, 2009, ISBN 978-1-60511-162-9

Volume 1190 — Active Polymers, K. Gall, T. Ikeda, P. Shastri, A. Lendlein, 2009,
ISBN 978-1-60511-163-6

Volume 1191 — Materials and Strategies for Lab-on-a-Chip — Biological Analysis, Cell-Material
Interfaces and Fluidic Assembly of Nanostructures, S. Murthy, H. Zeringue, S. Khan,
V. Ugaz, 2009, ISBN 978-1-60511-164-3

Volume 1192E —Materials and Devices for Flexible and Stretchable Electronics, S. Bauer, S.P. Lacour,
T. Li, T. Someya, 2009, ISBN 978-1-60511-165-0

Volume 1193 — Scientific Basis for Nuclear Waste Management XXXIII, B.E. Burakov, A.S. Aloy,
2009, ISBN 978-1-60511-166-7

Prior Materials Research Society Symposium Proceedings available by contacting Materials Research Society

# Low-k Dielectrics I

Mater. Res. Soc. Symp. Proc. Vol. 1156 © 2009 Materials Research Society
1156-D01-04

# Effect of Trapping on Dielectric Conduction Mechanisms of ULK/Cu Interconnects

V. Verrière[1,2], C. Guedj[2], D. Roy[1], S. Blonkowski[1], A. Sylvestre[3]
[1] STMicroelectronics 850 rue J. Monnet 38926 Crolles Cedex France
[2] CEA-Leti MINATEC 17 avenue des Martyrs 38054 Grenoble Cedex France
[3] Grenoble Electrical Engineering Lab, (G2ELab) CNRS 25 avenue des Martyrs BP166 38042 Grenoble Cedex 9 France

## ABSTRACT

Trapping in low-κ dielectric for interconnects was highlighted by voltage shift in IV current-voltage measurements. It is shown that effects of trapping can impact the extraction of conduction mechanisms. Capacitance measurements made on these materials reveal that trapping is at the origin in the increase of capacitance. The creation of dipoles because of this trapping explains this increase in the value of capacitance.

## INTRODUCTION

The drastic reduction of intra-level Metal-Metal spacing in advanced interconnects poses concern for reliability linked to the dielectric integrity. The Low-κ dielectric materials which compose the dielectric stack are the site of leakage currents under electric stress. These leakage currents damage the materials to the breakdown. The knowledge of the mechanisms linked to the leakage currents is a key to explain the damaging. Nevertheless the dielectric materials are composed of many defects, which can be active for conduction or just have a role of traps. Characterization of all these defects is an issue to establish the good diagnose of defectivity. Trapping had been already put in evidence in such structures [1]. We propose an analysis of trapping impact through leakage currents and capacitance measurements.

## EXPERIMENT

Test structures were fabricated with an advanced Cu/ Low-κ process with 45 nm node processes (Figure 1). Measurements were performed on comb-comb test structures (Figure 2). Leakage currents against field are measured by sweep IV with different speeds of voltage.

Dynamical behavior is studied by impedance spectroscopy for frequencies between $10^{-2}$ Hz and $10^3$Hz. A sinusoidal voltage $V_{rms}$=0.5 V is applied.

Measurements have been performed against temperature (between 100°C and 200°C).

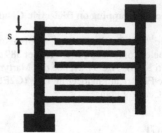

**Figure 1.** TEM cross sectional view of structures. SiCOH is porous with a porosity of about 30%.

**Figure 2.** Test structures are interdigited combs structures. Space s between lines is 70 nm.

## RESULTS AND DISCUSSION

### Measurement of leakage current against applied field and effect of trapping

Trapping takes place in virgin structures from the application of an electric stress. In this part its impact is studied during measurement of leakage currents. If two successive sweeps are performed, a voltage shift to high fields is observed between the two curves (Figure 3). On the other hand, the voltage speed has no influence on the value of the current.

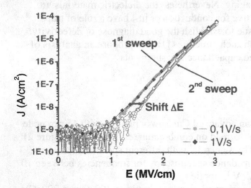

**Figure 3.** Leakage current measured by two successive sweeps at 125°C to 3MV/cm. Two sweep speeds have been tested: a fast one (1V/s) and a slow one (0.1V/s).

First it is observed that the voltage shift is observed from the base of the curves and decreases with increasing-field. At the base the shift $\Delta E$ is about 0.2 MV/cm. Nevertheless, the shape of the curves is strongly impacted after trapping and this will compromise the extraction of the conduction mechanism.

4

Poole-Frenkel conduction model is detected from the first sweep with the following field-dependence:

$$J(E) \sim \exp\left(-\frac{\phi_{PF}}{k_B T} + \frac{\beta_{PF}\sqrt{E}}{k_B T}\right) \text{ (Eq. 1)}$$

where $\Phi_{PF}$ is the activation energy linked to Poole-Frenkel effect and $\beta_{PF}$ the Poole-Frenkel defined by $\beta_{PF} = \sqrt{\dfrac{q}{\pi \varepsilon_0 \varepsilon_r}}$, with q the electron charge, $\varepsilon_0$ the vaccum permittivity and $\varepsilon_r$ the relative permittivity.

A permittivity of $\varepsilon_r$ between 2.5 and 3 is obtained from the fit.

The voltage shift corresponds to the internal field introduced by a uniform density of trapped charges $n_t$ and given by:

$$E_{trap} = -\frac{q n_t}{2\varepsilon_0 \varepsilon_r} \text{ (Eq. 2)}$$

During the measurement, $E_{trap}$ increases whereas the applied field increases. The measured conduction current settled according to the total field $E+E_{trap}$ and consequently according to the establishment speed of the total field. Therefore the establishment speed of the total field is a competition between establishment of the applied field and the internal field due to trapped charges. The measured current settles according to the faster establishment.

Thus the second curves in figure 3 correspond to partially trapped states, and since the applied field settles faster than the internal field, the curve is distorted.

Measurements have been performed to higher fields (Figure 4).

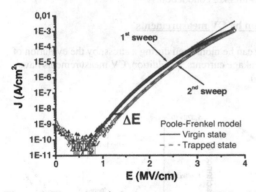

**Figure 4.** Two successive measurements at 125°C to high fields (about 4MV/cm).

At about 3MV/cm, the first sweep diverges from the Poole-Frenkel conduction model. Moreover, the Poole-Frenkel model is recovered at the second sweep, with a voltage shift. On the first sweep, the divergence from Poole-Frenkel conduction model takes place when the internal field due to trapped charges settles faster than the applied field [2], since it opposes. The recovering of Poole-Frenkel model at the second sweep means that most of the traps have been filled. We can extract the total traps density from the voltage shift and Eq. 2, at about $n_t \approx 5.10^{11} cm^{-2}$. Since all traps are filled, measurements of leakage currents are not anymore impacted by trapping. The conduction mechanisms can be studied against temperature (Figure 5).

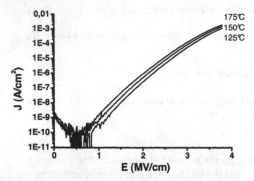

**Figure 5.** Leakage currents against temperature at trapped state to 4MV/cm.

Conduction currents correspond to Poole-Frenkel conduction with an activation energy $\Phi_{PF}$ of 0.8 eV.

Consequently, two types of defects have been set in evidenced: traps which are filling during electric stress and defects active for Poole-Frenkel conduction.

### Trapping during stress: characterization by CV measurements

From the previous observations, trapping can be monitored during a stress, by the evolution of voltage shift between measurements of leakage currents. In addition, CV measurements have been performed simultaneously (Figure 6).

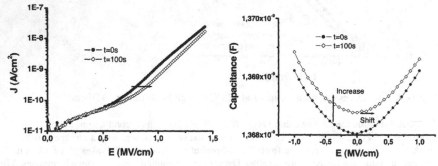

**Figure 6.** Leakage currents and capacitance (at 1 kHz) measured at 125 °C during a stress at

constant voltage (10V i.e. 1.43 MV/cm).

First it is observed the same voltage shift both in IV curves and CV curves. Moreover it is observed the increase of the capacitance and the decrease of the curvature of the parable. Consider a medium which contains $N$ dipoles of polarizability $\alpha$ and dipole moment $\mu$, with the assumption of low density so that interactions between dipoles are neglected. In the most general case concerning the intrinsic properties of the medium, the permittivity $\varepsilon$ against applied field is [3]:

$$\varepsilon(E) = \varepsilon_r(E,T) + N\left(\alpha + \frac{\mu^2}{3k_BT} - (1 - 4u - 2u^2)\frac{\mu^4 E^2}{45k_B^3T^3}\right)$$

where $u$ is the term for anisotropic polarization linked to the polarization of the $N$ dipoles. In the case of isotropy, the permittivity becomes:

$$\varepsilon(E) = \varepsilon_r(E,T) + N\left(\alpha + \frac{\mu^2}{3k_BT} - \frac{\mu^4 E^2}{45k_B^3T^3}\right) \text{ (Eq. 3)}$$

$\varepsilon_r(E,T)$ is the intrinsic permittivity of the medium.

Equation 3 allows us to speculate that the increases of the capacitance and the decrease of the curvature can be explained by the addition of a low density of $N_{add}$ dipoles of dipole moment $\mu_{add}$, whom the polarizability is neglected, compared to the steady-state dipole. With this assumption, the total permittivity according to the initial permittivity $\varepsilon_{virg}$ of the virgin dielectric stack (measured before stress at t=0s) and the addition of $N_{add}$ dipoles is given by:

$$\varepsilon(E) = \varepsilon_{virg}(E,T) + N_{add}\left(\frac{\mu_{add}^2}{3k_BT} - \frac{\mu_{add}^4 E^2}{45k_B^3T^3}\right) \text{ (Eq. 4)}$$

A dipole moment $\mu_{add}$ of about 35 Debye is extracted using Eq. 4. In the case of a dipole between charges e and –e, the length of the dipole is about 8 Å. Then we obtain $N_{add}$ of the order of $10^{16}$ cm$^{-3}$. With the assumption that dipoles were uniformly distributed between Cu lines, we extract the corresponding surface density, to be compared to the density of filled traps, at the same time during the stress (Figure 7).

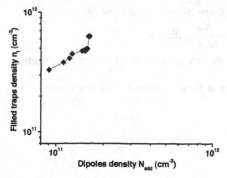

**Figure 7.** Correlation between dipoles density and filled traps density during stress at constant voltage (1.43MV/cm) and 125°C.

Surface densities of created dipoles and filled traps are well correlated. Thus capacitance can be used to monitor trapping during stress. From this observation, we proposed to characterize the state during stress, by measurement of capacitance versus frequency with zero bias.

## Dynamic behavior by impedance spectroscopy and monitoring of stress

Since we have linked increase of capacitance with trapping, we have characterized and studied the stress by measurement of the capacitance against frequency non-destructively. To perform this, the impedance spectroscopy allows probing the dynamic behavior without bias on a large range of frequencies.

First we have characterized the capacitance at virgin state (Figure 8) before any stress.

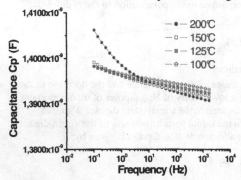

**Figure 8.** Capacitance against frequency and temperature at virgin state (before stress).

For each temperature, capacitance increases at low frequencies. At high frequencies, capacitance is decreasing with increasing temperature, and at low frequencies, the temperature-dependence is opposing. At high frequencies the temperature-dependence corresponds to the dipolar contribution of dipoles.

With the same assumption established for Eq.3, the permittivity can be written:

$$\varepsilon_S(T) = \varepsilon_r(T) + \frac{N\mu^2}{3\varepsilon_0 k_B T} \text{ (Eq. 5)}$$

$\varepsilon_r(T)$ is the intrinsic permittivity at zero bias, N the density of dipoles present at low density in the virgin stack and $\mu$ the corresponding moment.

Capacitance has been fitted according to Eq. 5, with the assumption that the temperature-dependence of $\varepsilon_r(T)$ was neglected (Figure 9).

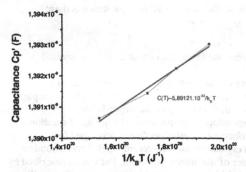

**Figure 9.** Capacitance against temperature at 1kHz.

From Eq. 5, we obtain $N\mu^2 \approx 5.10^{-33}$ $C^2m^{-1}$ that corresponds to a density $N \approx 6.10^{20}$ cm$^{-3}$ for a dipole moment of around 1 Debye unit. From this, we can estimate dipole-dipole interaction energy given by $E_{dip} = \dfrac{\mu^2}{4\pi\varepsilon_0 d^3}$ with $N \approx \dfrac{1}{d^3}$, with d is the dipole-dipole space, and then approximately

$E_{dip} \approx \dfrac{N\mu^2}{4\pi\varepsilon_0}$. We obtain $E_{dip} \approx 0.3$meV that is much lower than the thermal energy $E_{th} \approx 25$meV at 300K. Thus the assumption of a low density of dipoles in the dielectrics at virgin state is well verified.

Concerning the capacitance in low frequencies, the increase to low frequencies has not yet been studied in details, but has to be linked to the contribution of hopping of charges carriers in amorphous dielectrics [4].

From this, the capacitance measured by impedance spectroscopy has been used to characterize the state during a stress.

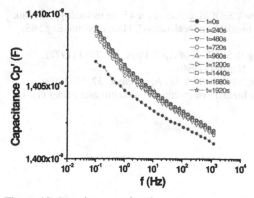

**Figure 10.** Capacitance against frequency measured by impedance spectroscopy at 125°C and during a stress at 1.43MV/cm.

As previously capacitance increases. Using Eq. 4, the increase $\Delta C$ of capacitance is then:

$$\Delta C = \frac{N_{add} \mu_{add}^2}{3 k_B T} \frac{\Sigma}{s}$$

On figure 10, at 1920s from the start of the stress, we can extract about $2 \ 10^{11} \text{cm}^{-2}$ created dipoles.

Concerning the physical mechanism explaining the link between trapping and creation of dipoles, model has been already proposed in the case of charge trapping in $SiO_2$ [5]. A simple model describes the increase of the polarization in the vicinity of the trapped charge, by interaction of the trapped charge with the components of the medium. The model of creation of dipoles by charges trapping is similar to the concept of crystalline dipole developed for crystalline defects [3]. As it is described for a crystalline dipole, created dipoles correspond to charges e and –e spaced by a length of the order of magnitude of a bond. The model described by Eq. 4 applies the Langevin-Debye model that is likewise recovered in the concept of crystalline dipole. It highlights the relevance of capacitance as a monitor of trapping.

## CONCLUSIONS

The effect of trapping has been studied on leakage currents and on capacitance. It has been shown that during trapping, measurements of leakage currents against applied field are strongly modified by the local field created by trapped charges. The extraction of conduction mechanisms is thus compromised. The trapping has been linked to the increase of capacitance measured by CV. With the assumption of low density, creation of dipoles has been proposed as interpretation to explain the increase of capacitance. The good correlation obtained between dipoles densities and filled traps densities show that capacitance is a good monitoring for trapping. This justifies the use of impedance spectroscopy as a non-destructive technique for characterization of trapping. Moreover this technique is promising to probe the behavior of the dielectric stack, non-destructively and close to use conditions, i.e. at low fields.

## REFERENCES
1. M. Vilmay et al., "Characterization of low-k SiOCH dielectric for 45 nm technology and link between the dominant leakage path and the breakdown localization", Microelectron. Eng., **85**, 2075 (2008).
2. P. Solomon, "High-field electron trapping in $SiO_2$", J. of Appl. Phys., **48**, 3843 (1977).
3. R. Coelho, "Propriétés des diélectriques", Hermès (1990).
4. A.K. Jonscher, "Dielectric relaxation in solids", J. Phys. D.: Appl. Phys., **32**, 57 (1999).
5. G. Kamoulakos et al., "Unified model for breakdown in thin and ultrathin gate oxides (12-15 nm)", J. of Appl. Phys., **86**, 5131 (1999).

Mater. Res. Soc. Symp. Proc. Vol. 1156 © 2009 Materials Research Society     1156-D01-05

# Dual Damascene Reactive Ion Etch Polymer Characterization Through X-ray Photoelectron Spectroscopy for 65 nm and 45nm Technology Nodes

Samuel S. Choi, Chet Dziobkowski[1], Leo Tai[1]

IBM TJ Watson Research Center, 2070 Route 52, Hopewell Junction, NY 12533, U.S.A.
[2] IBM Microelectronics SRDC, 2070 Route 52, Hopewell Junction, NY 12533, U.S.A.

## ABSTRACT

At 65nm and beyond technology nodes, copper interconnect formation in dual damascene integration is continually challenged from a polymer management perspective. Highly polymeric plasma chemistry is required to reduce line edge roughness, shape physical profile, and control critical dimension across a 300mm wafer. But too much fluorocarbon deposition on a wafer results in poor defects yield.

In this paper, X-ray photoelectron spectroscopy (XPS) characterization technique is used to quantify and to optimize a metal line reactive ion etch process to increase electrical opens yield. A reduction of 2 at.% in carbon mass results in a Do (defects/cm$^2$) improvement from > 2.0 to less than 1.0. This result is realized without a shift to the trench physical profile which is important for reliability performance. Moreover, with a shorter turnaround time of XPS characterization compared to electrical hardware splits, quicker yield learning cycle is realized for both RIE process and module integration.

## INTRODUCTION

Copper interconnect process development and manufacturing are extremely challenged as technology node's critical dimension (CD) shrinks and chip functional yields must be met to profit [1]. Much effort is devoted to RIE pattern transfer to meet each technology nodes' critical dimension specifications. Fluorocarbon species are extensively used in RIE to shrink and to control CD [2]. A consequence of too much fluorocarbon (e.g. $CF_x$) on a wafer is manifested in a Photo-Limiting Yield (PLY) scan detection of hollow metal, Fig. 1. Hollow metal is a killer defect that degrades electrical open yield.

Figure 1: Corroded copper lines due to residual RIE polymer.

Unfortunately, much less effort is devoted for manufacturability of a RIE process in relation to electrical chip yield. Namely, how can a RIE process be optimized such that both technology and electrical yield requirements are met? This paper addresses this need. Furthermore, new and rapid turnaround characterization techniques are in need to optimize a RIE process that is scalable to the next technology node [3].

We propose the use of X-ray photo-electron spectroscopy (XPS) surface analytic technique as an additive tool for RIE process optimization. The XPS's soft x-ray energy is less energetic than an electron beam, hence mono-layers of fluorine detection is possible. Moreover, unlike Auger electron spectroscopy, XPS's large probe beam area ($10 \ \mu m^2$) provides ample sampling desirable for 300 mm wafers [4]. Standard scanning electron micrograph (SEM) analysis is still used to confirm physical profile shifts associated with RIE process changes. Lastly, electrical split of 65nm RIE processes shows enhanced electrical open yield that correlates to reduced fluorocarbon content. A parameter defined as Do, measured defects per square area, is used to predict chip yield, Y, in a Poisson's model, $Y = e^{-Do * Ac(p)}$ [5]. The parameter Do is used to assess 65nm RIE process optimization improvements in this paper. And case study of how to optimize 45nm metal line process is discussed.

## EXPERIMENT AND DISCUSSION

The 65nm 1× metal line level of the dual damascene integration is investigated. Figure 2 shows a schematic of the trench integration film stack in which films 5, 4, 3, 1, and 2 are etched sequentially [3]. Note, after film 1 is etched, in-situ ash is utilized to etch the organic mask of film 3 which leaves an exposed dielectric surface to film 2's plasma condition, the final etch step. Figure 3 shows the completed pattern transfer profile. All etches are conducted on a TEL JIN chamber equipped with 60/13.5 MHz RF powers respectively. After etch, wafers are transferred to a XPS chamber for analysis [4]. All samples for XPS analysis are not subjected to hydrofluoric acid treatment post RIE.

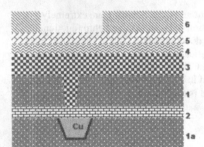
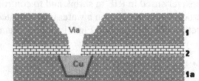

Figure 2: Films are 1. low-k dielectric, 2. SiN, 3.organic, 4. oxide, 5. BARC 6. resist

Figure 3: Final metal/via profile post RIE pattern transfer

The cap etch step (film 2 in Fig.2) is selected for change since a wafer's entire dielectric surface is exposed to polymeric deposition. The cap etch plasma condition is modified to reduce fluorocarbon generation. This is accomplished by a 50% reduction in plasma source power while pressure and gas flows are maintained. Two wafers, wafer A represents the baseline RIE process and wafer B denotes the 50% reduction in plasma source power, are probed with XPS.

Two sites on a 300mm wafer are probed, wafer center and edge. The results from the two locations are consistent so only the wafer center data is shown. Figures 4 and 5 are the XPS elemental spectra of the two processes A and B, respectively. Silicon and oxygen peaks are from the dielectric film. The detection of both fluorine and carbon is indicative of fluorocarbon residues on a wafer's surface.

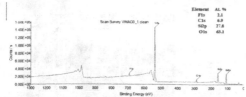

Figure 4: Process A has higher carbon atomic mass content of 6.9%

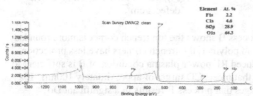

Figure 5: Process B with 50% reduced source power has a carbon atomic mass of 4.6 at.% which is 2% lower than process A.

Table 1 shows two rows of data for each process A and B. The "as is" row is the carbon content probed after RIE. The "sputtered" row data is the carbon content after an argon ion beam sputtering (<1nm thick) of the sample surface to remove native oxide. The resultant carbon content differences of the two rows are consistent in which process B is approximately 2.at% lower. Note the detection of fluorine for the "sputtered" condition.

| Process/ elements | C1s | F1s | Si2p | O1s |
|---|---|---|---|---|
| RIE A (as is) | 9.3 | 2.2 | 27.2 | 61.4 |
| RIE A (sputtered) | 6.9 | 2.1 | 27.8 | 63.1 |
| RIE B (as is) | 7.8 | 2.0 | 27.8 | 63.1 |
| RIE B (sputtered) | 4.6 | 2.2 | 28.9 | 64.3 |

Table 1: XPS elemental (C1s carbon, F1s fluorine, Si2p silicon, O1s oxygen) spectra for 65nm Mx process A and B. Process B shows a lower carbon atomic percent.

To assess the impact of 2 at.% of carbon mass reduction on electrical yield, a split of 5 wafers each for process A and B are compared. There is no difference in electrical shorts for the two processes. However, for electrical opens, process A exhibited a Do value ranged from 2-3 defects/cm$^2$ whereas process B the averaged Do is less than 1.0 defects/cm$^2$. This is a significant difference with respect to the exponential dependence of Do on chip yield described in the introduction.

SEM images of process A and B are shown in Figure 6 and 7, respectively. The pitch of the nested lines is 200 nm.

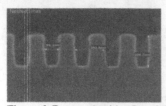

Figure 6: Process A with a Do ~3.0 defects/cm$^2$.

Figure 7: Process B with Do <1.0 defects/cm$^2$.

To be critical, SEM image from process B shows the top trench corner is more rounded than process A. Due the reduction in polymer, the trench corners have less protection against incoming ions. But the reduced RF power plasma condition of B is sufficient to etch a comparable physical profile that meets CD target of approximately 100nm. It's not obvious from the two SEM images that process B results in a significant Do improvement. Without the XPS and Do results, RIE process development ends with process A. Consequently, a difficult path for module yield improvement lies ahead.

At the 45nm node, a baseline metal line RIE process C is arbitrarily defined to meet the technology's physical specifications and has a fluorocarbon content of 4.55 at.% as shown in Table 2. The objective is to optimize process C in terms of reduced fluorocarbon with a minimal change to the trench physical profile. Select plasma conditions of process C are explored and then characterized with XPS and SEM.

| 45nm Mx Process /XPS | XPS "C" at% | XPS "F" at% |
|---|---|---|
| [C] Baseline Process | 4.55 | 1.36 |
| [F] O$_2$ only Ash | 3.28 | 1.38 |
| [G] CO$_2$+O$_2$ mix Ash | 3.53 | 1.76 |
| [D] Cap etch w/ O$_2$ gas | 2.93 | 1.75 |
| [E] High Pressure CO$_2$ Ash | 2.66 | 1.85 |

Table 2: 45nm Mx process variations and the respective XPS carbon and fluorine contents. Fluorine detected for all processes from the cap etch step.

The in-situ ash and cap etch steps' plasma condition is adjusted independently, films 2 and 3 in Fig.1. Table 2 list the key plasma condition parameter examined. Three different ash conditions are examined in which the baseline process uses carbon dioxide CO$_2$ gas only. Oxygen (O$_2$) gas effectively removes fluorocarbon species, but also damages low-k (k <3.0) dielectric film, e.g. CD blowout. But at low pressures (<50 mT increase anisotropic etch) and RF power (<1000W 13.5MHz) plasma conditions, low-k dielectric damage is minimized. Three ash regimes are examined, process E is the high pressure (>100mT) $CO_2$ ash, process F is the low pressure/power $O_2$ ash, and process G is a balance to reduce low-k damage and to reduce fluorocarbon. All of the ash etch times are optically end pointed. Lastly, typical cap etch chemistries incorporate fluorine, hydrogen, carbon dioxide, and argon gases. Process D examines the incorporation of $O_2$ gas in the aforementioned gases.

The last 2 columns Table 2 shows the XPS carbon and fluorine content for the respective processes, C-G. The arbitrary baseline process C has the highest carbon content, 4.55 at.%. Process F and G have 1 to 1.2 at.% lower carbon content than C. The high pressure process E exhibits the least amount of fluorocarbon. Process D, cap etch step with $O_2$ gas, also exhibits a comparable carbon content of 2.93 at.% as process E. From these XPS results, ash conditions greatly affect a wafer's final surface carbon content independent of the next cap etch step. This suggests that the baseline ash process ineffectively removes residual organic film or carbon from the $CO_2$ ash plasma remains on a wafer after the cap etch step. Given that the carbon reduction has been demonstrated by these processes, the impact of the process change to physical profile is required.

Cross section SEM images are obtained for each of the processes discussed above. Figure 8 is the trench profile of the baseline process C. SEM images of processes D, E, F, and G are compared to Fig. 8. Figure 9 of process D shows comparable physical profile with 1.5 at.% reduction in carbon. This indicates appropriate introduction of $O_2$ gas in plasma steps is critical to remove carbon. Figure 10 is the high pressure $CO_2$ ash. Here, the lateral etch component exceeds polymer deposition rate and the reduced surface area resulted in the lowest carbon detection. Processes F and G, Fig. 11 and 12 (both 1 at.% lower carbon), exhibit a contrast in physical profile. Process G's $CO_2/O_2$ plasma condition results in a faster lateral etch component that rounds corners, but at reduced pressure to process E. The $O_2$ ash of process F effectively removes carbon like Process G without rounded corners due to the low pressure/power. At a low pressure, an ion's mean free path is increased, contrast to processes E, F. For low-k films, a high degree of etch anisotropy is required with $O_2$ plasmas. And low power results in reduced trench corner facets. These results show that process windows exist in which both technology specification and yield improvement are possible. Lastly, for all processes except E, subsequent metallization and chemical mechanical polish, the final metal profile become all comparable. Unfortunately, the 45nm Mx electrical Do values are not available at the time of this paper.

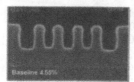

Figure 8: Baseline Process C

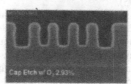

Figure 9: Process D (cap etch with $O_2$)

15

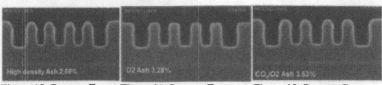

Figure 10: Process E          Figure 11: Process F          Figure 12: Process G
(High Pressure $CO_2$)        ($O_2$ only Ash)               ($CO_2/O_2$ ash)

## CONCLUSION

At 65nm, significant defect density, Do, improvement is realized with a reduced fluorocarbon content RIE process as detected by XPS microprobe. A reduction in 2 at.% of carbon resulted in a Do electrical open improvement from 3 to less than 1.0 defect/$cm^2$. There is little change to the physical profile as seen in SEM images, Figs. 6&7. Likewise for 45nm, the cap etch and the in-situ ash steps are explored to reduce fluorocarbon content. The cap etch step with $O_2$ addition reduced carbon content by 1.5 at% with no impact to the trench physical profile. RIE processes are optimized to reduce carbon content on a wafer with XPS microprobe's ability to detect fluorocarbon species. Electrical opens yield enhancement is not realized through SEM image characterization only.

For future work, XPS microprobe's ability to distinguish polymer thickness and density on a wafer field versus a trench side wall may become important especially for lower k dielectric material with exotic hard mask integration.

## ACKNOWLEDGMENTS

Authors would like to thank the following IBM personnel for support of this work Christine Bunke, Thomas Houghton, Kaushik Kumar, Dan Edelstein, Robert Wisnieff, and Thomas Ivers.

## REFERENCES

1. D. C. Edelstein, G. A. Sai-Halasz, Y.-J. Mii, "VLSI on-chip interconnection performance simulations and measurements," IBM J. Research and Development Vol. 39, No. 4 (1995).
2. T. J. Dalton, N. Fuller, G. Gibson, K. Kumar, "Dielectric Etching Technology Review" American Vacuum Society Conference PS-TuA1 4 Nov (2003).
3. K. Kumar, "Plasma-Lithography Interactions for Advanced CMOS Manufacturing (45nm and Beyond)" AVS, October 2008.
4. Tool description and fundamentals of XPS http://www.phi.com/techniques/xps.html
5. Dennis J. Ciplickas, et al. "Advanced Yield Learning Through Predictive Micro-Yield Modeling" Proc. 7th IEEE/ISSM, Oct. 1998.

Mater. Res. Soc. Symp. Proc. Vol. 1156 © 2009 Materials Research Society       1156-D01-06

# Interaction of O and H Atoms With Low-$k$ SiOCH Films Pretreated in He Plasma

O. V. Braginsky, A. S. Kovalev, D. V. Lopaev, Y. A. Mankelevich, E. M. Malykhin, O. V. Proshina, T. V. Rakhimova, A. T. Rakhimov, A. N. Vasilieva, D. G. Voloshin, S. M. Zyryanov, M. R. Baklanov[1]

Skobeltsyn Institute of Nuclear Physics, Lomonosov Moscow State University, Moscow, Russia
[1] IMEC, Leuven, Belgium

## ABSTRACT

The effect of He plasma pretreatment on the interaction of O and H atoms with SiCOH low-k materials is studied using a special experimental system designed for this purpose. The experimental system allowed separate studies of the effects of He plasma, VUV light and He $2^1 S_0$ metastable atoms. It is shown that carbon depletion by oxygen atoms can be significantly reduced by He plasma pretreatment. Considerable increase of CH and $CH_2$-$CH_2$ groups in the surface area of low-k films is observed when the films were exposed to VUV light and metastable atoms generated by He plasma. FTIR and ellipsometry showed the formation of a densified surface layer. This carbon rich densified surface layer decreases damage of the low-k film when it is exposed to $O_2$ plasma. The impact of H atoms on low-k surfaces noticeably differs from O atoms effect. The H atoms saturate all unbonded remaining carbon bonds thereby promoting improvement of SiOCH structure.

## INTRODUCTION

It is known that pretreatment of SiCOH low-k materials in He plasma afterglow significantly reduces the plasma damage during the following exposure in $NH_3$ and $O_2$ plasma. It was speculated that VUV photons formed in He plasma are able to modify the surface of the low-k material, create active surface centers that stimulate recombination (deactivation) of active radicals formed in the strip plasma [1]. The recombination decreases the number of active radicals penetrating into the pores of low-k materials. The experimental technique used for the loss probability study and the results of analysis are described in our previous publication [2]. It was shown that the loss probabilities of O and H radicals are described in the frames of Langmuir–Hinshelwood theory of heterogeneous recombination of physically adsorbed atoms (PO sites with bond enthalpy D(P-O)~0.2 eV) and chemisorbed atoms (SO sites with bond enthalpy D(S-O)~1.6 eV). It was concluded that the surface recombination is the main mechanism of the O and H atoms loss. The reaction of these atoms with the carbon containing groups has a secondary order contribution. Therefore, the surface recombination defines the profile of the radicals concentration in low-k films and, therefore, the depth of plasma damage.

The atoms penetrate into the pores due to random walk mechanism. The measured recombination coefficients allowed us to estimate the typical penetration depth ($L_{pen}$~24 nm) of O atom into the studied low-k materials with pore diameter close to 2 nm. This depth is in reasonable agreement with the experimentally determined depth of the $CH_3$ groups depletion.

Using an approach based on comparison of $SiCH_3$ reactions with so called "gas phase analogues" with known reaction rate constants, it was concluded that the reaction of $SiCH_3$ groups with O and H atoms starts with detachment of one hydrogen atom [2]:

$$SiCH_3 + O \rightarrow SiCH_2 + OH$$
$$SiCH_3 + H \rightarrow SiCH_2 + H_2$$

Further transformations of $SiCH_2$ groups depend on ambient. If the low-k films are located in $O_2$ plasma, further reactions lead to complete degradation (hydrophilization):

$$SiCH_2 + O \rightarrow Si + CH_2O$$
$$Si + O \rightarrow SiO$$
$$SiCH_3 + SiO \rightarrow SiCH_2 + SiOH$$
$$...$$

The situation is completely different in the case when the low-k material is located in $H_2$ plasma. In this case further interaction of $SiCH_2$ groups with H radicals leads to restoration of initial Si-CH$_3$ groups. Therefore, technological processes based on hydrogen radicals (for instance, hydrogen plasma afterglow) with noble carrier gases are promising candidates for damage free resist strip in the low-k based technology. This conclusion is supported by the results demonstrating damage free resist strip in He/H$_2$ plasma afterglow [3] and by the results of SIMS analysis reported by Lazzetti [4]. In the last paper, the low-k films were exposed in $D_2$ plasma and it is shown that $SiCH_3$ groups are loosing part of hydrogen atoms but they are immediately saturated by D-atoms. Finally, the sum of H and D is equal to the initial concentration of hydrogen atoms in $SiCH_3$ groups.

In this work, interaction of O and H radicals with low-k films treated in He plasma is studied. The reactor used for He plasma treatment allowed to separate effects of VUV light, He $2^1S_0$ metastable atoms (He$_m$ hereafter) and ions, and to monitor and control the plasma characteristics. The SiCOH samples were analyzed using FTIR and ellipsometry. The main purpose of this work is to get additional information necessary to understand mechanisms of damage reduction when the low-k materials are exposed to He plasma.

## EXPERIMENTAL DETAILS AND DISCUSSION

Three porogen based SiCOH low-$k$ films (CVD1, CVD2, CVD3) were deposited by plasma enhanced chemical vapor deposition (Table I). A near monochromatic UV lamp with $\lambda = 172$ ($\pm 15$) nm was used to cure CVD2 and CVD3 and a broadband lamp with $\lambda > 200$ nm to cure the film CVD 1. Experimental set-up and technique for measurements of O and H atom loss probabilities were described in detail earlier [2].

**Table I.** The studied low-k SiOCH materials.

| Property | CVD1 | CVD2 | CVD3 |
|---|---|---|---|
| Porosity (%) | 24 | 24 | 33 |
| Pore radius (nm) | 0.8 | 0.8 | 0.9-1.0 |

He plasma was generated in low-pressure rf discharge (80 MHz) in flow of pure He at 20 mTorr in a long quartz tube with inner diameter of 55 mm (Fig.1). The system allowed obtaining a long plasma column with slowly decreasing electron temperature and plasma potential, and gradually decreasing plasma density. The plasma parameters along the tube were measured by a moveable Langmuir probe. To understand the He plasma effect on SiOCH films the samples received different treatments in He plasma. The samples 5 and 4 are located directly in the plasma column. Sample 5 was closer to the rf antenna in a location of higher plasma density, whereas plasma potentials and energy of ions falling to samples 5 and 4 were approximately the same. Sample 3 was in the far plasma afterglow but faced the plasma column and, therefore, irradiated by intensive VUV emission. Sample 1 was also in the afterglow but was turned to the opposite side of the plasma column and only metastable $He_m$ atoms were able to reach the SiOCH surface.

The O and H loss probabilities ($\gamma_O$ and $\gamma_H$, respectively) measured in 13.56 MHz discharge afterglow are presented in Tables II and III. All these experiments were repeated 4 - 5 times and the measurement statistics is presented as an error in Table II. The samples pretreatment in He plasma leads to noticeable decrease of $\gamma_O$ values. The $\gamma_H$ values do not show significant changes after He plasma treatment. By using a moveable Langmuir probe, plasma density, plasma potential and electron temperature were determined from the probe VI characteristics as function of position along the plasma column at the different discharge powers. The measured plasma density in He discharge at 25W inputted power is shown in Fig. 2.

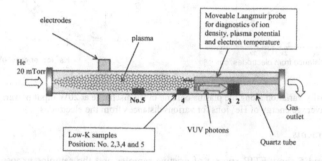

**Fig.1.** The experimental set up for pretreatment of SiOCH samples in He plasma.

Table II. Loss probabilities for O atoms $\gamma_O$ ($\delta\gamma_O = \pm 0.5 \cdot 10^{-3}$)

| Low-K sample | Non-treated samples + O atoms); # 1 | He plasma (no direct VUV, only $He_m$); # 2 | He plasma afterglow (VUV + $He_m$; #3 | He plasma ($N_i \approx 10^{10}$ cm$^{-3}$, $\varepsilon_i \approx 23\pm5$ eV), #4 | He plasma ($N_i \approx 5 \cdot 10^{10}$ cm$^{-3}$, $\varepsilon_i \approx 28\pm5$ eV); # 5 |
|---|---|---|---|---|---|
| CVD1 | $4.6 \cdot 10^{-3}$ | $3.6 \cdot 10^{-3}$ | $6.4 \cdot 10^{-3}$ | $2.4 \cdot 10^{-3}$ | $1.2 \cdot 10^{-3}$ |
| CVD2 | $2.6 \cdot 10^{-3}$ | $4.2 \cdot 10^{-3}$ | $3.1 \cdot 10^{-3}$ | $1.6 \cdot 10^{-3}$ | $1 \cdot 10^{-3}$ |
| CVD3 | $2.8 \cdot 10^{-3}$ | $2.6 \cdot 10^{-3}$ | $2.4 \cdot 10^{-3}$ | $2.2 \cdot 10^{-3}$ | $1.6 \cdot 10^{-3}$ |

Table III. Loss probabilities for hydrogen atoms $\gamma_H$ ( $\delta\gamma_O = \pm 1 \cdot 10^{-4}$)

| Low-K sample | Non-treated samples + O atoms); # 1 | He plasma (no VUV, only $He_m$); #2 | He plasma afterglow (VUV + $He_m$) # 3 | He plasma ($N_i \approx 10^{10}$ cm$^{-3}$, $\varepsilon_i \approx 23\pm5$ eV), #4 | He plasma ($N_i \approx 5 \cdot 10^{10}$ cm$^{-3}$, $\varepsilon_i \approx 28\pm5$ eV); # 5 |
|---|---|---|---|---|---|
| CVD1 | $8 \cdot 10^{-4}$ | $4 \cdot 10^{-4}$ | $3.4 \cdot 10^{-4}$ | $3.3 \cdot 10^{-4}$ | $5 \cdot 10^{-4}$ |
| CVD3 | $3 \cdot 10^{-4}$ | $2 \cdot 10^{-4}$ | $2.4 \cdot 10^{-4}$ | $2.2 \cdot 10^{-4}$ | $2.5 \cdot 10^{-4}$ |

To estimate the ion energy distribution function (IEDF) on the samples, the ion motion in the given radial potential with the ambipolar field $E_r$ was simulated by Monte Carlo method. The calculated energy spectra of He$^+$ ions at different distances from the electrodes are shown in Fig.3. It can be seen that IEDF is rather wide and slightly changes with the distance along the tube towards the end of the plasma column. These small changes in contrast to the large variations of ion density and ion flux to the samples apparently will lead approximately to the same ion effect on surface in He plasma. Therefore, we will be able to assume that the ion energy is nearly constant along the tube.

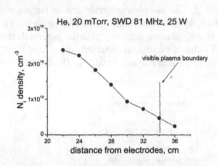

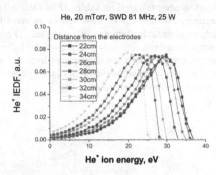

Fig.2 Plasma density variation along the plasma column in He discharge at 20W input power.
Fig.3 IEDF and average energy of He$^+$ ions for various distances from the electrodes

## FTIR measurements

Figures 4 and 5 show FTIR spectra of pristine samples and the samples exposed to oxygen and hydrogen plasmas. Analysis of FTIR data allow to make the following qualitative conclusions:
1. Exposure of all studied pristine low-k materials in oxygen plasma leads to
- carbon depletion (both C from (Si-CH$_3$) and C from hydrocarbons residue (–C-C-H frame);
- significant hydrophilization and water adsorption (peaks between 3000 and 3500 cm$^{-1}$);
- linear Si-O-Si structure transforms to the ordered network Si-O-Si similar to usual SiO$_2$ (1000-1200 cm$^{-1}$).
2. Modification of these materials during the exposure in hydrogen plasma is not significant. The most pronounced change is related to formation of C-C=O and CHx groups (peaks at 1700-1800 cm$^{-1}$ and 2800-3000 cm$^{-1}$, respectively). SiCH$_3$ groups depletion and hydrophilization are not observed.
3. Exposure of low-k films in VUV photons plasma decreases the degree of hydrophilization during the following exposure in oxygen plasma. This effect becomes even stronger after exposure in He plasma. No hydrophilization is observed in this case. However, effects of "de-hydrophilisation" seem different for CVD1 and CVD2. A strong increase of CHx groups concentration and formation of CH$_2$-CH$_2$ groups is observed in CVD1, while these changes are not visible in CVD2.

The well-defined changes in the baseline of reflection and transmission in the FTIR spectra of CVD1 at high wavenumbers are related to the reduction of film transmission because of the formation of a surface layer having different optical characteristics. This is related to surface densification. Therefore, the effect of O plasma on the low-K surface should be reduced because of "surface sealing". Similar behavior of CVD1 film was observed in the paper [3]. The formation of a dense surface layer and pore sealing was proven by X-ray reflectivity measurements and by ellipsometric porosimetry. Different behavior of CVD2 has also been shown [5]. The sealing of these films requires much higher energy of He ions. This is the reason why the transmission reduction is not observed in our conditions. Surprisingly, we observed that this dense layer is mainly formed by hydrocarbons. The origin of the hydrocarbon compounds is not completely clear. One can assume that these hydrocarbons form as a result of modification of amorphous carbon like porogen residues observed in all CVD based SiCOH films [6].

4. The film CVD3 is from the same family as CVD2. Probably, this is the reason why the formation of a dense surface layer during the exposure in $H_2$-plasma is not as pronounced as in the case of oxygen plasma (see the region 5000-7000 $cm^{-1}$). However, exposure of CVD3 in hydrogen plasma leads to formation of CHx and $CH_2$-$CH_2$ while they are not formed in CVD2 exposed in oxygen plasma. The formation of a dense layer is in agreement with our previous data [3].

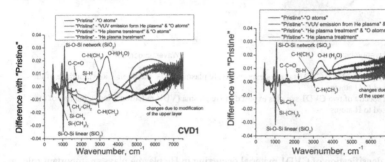

**Fig.4a.** Differential FTIR spectra of pristine CVD1 sample and the samples exposed to 1) O atoms; 2) VUV from He plasma followed by O atoms; 3) He plasma followed by O atoms; 4) only He plasma.
**Fig.4b.** Differential FTIR spectra of pristine CVD2 sample and the samples exposed to 1) O atoms; 2) VUV from He plasma followed by O atoms; 3) He plasma followed by O atoms; 4) only He plasma.

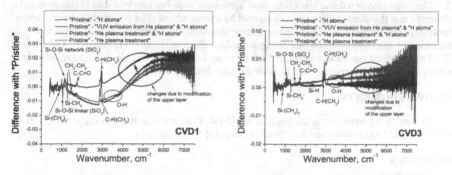

**Fig.5a.** Differential FTIR spectra of pristine CVD1 sample and the samples exposed to 1) H atoms; 2) VUV from He plasma followed by H atoms; 3) He plasma followed by H atoms; 4) He plasma.
**Fig.5b.** Differential FTIR spectra of pristine CVD3 sample and the samples exposed to 1) H atoms; 2) VUV from He plasma followed by H atoms; 3) He plasma followed by H atoms; 4) He plasma.

Figures 6 and 7 show differential FTIR spectra of CVD1 films exposed to He plasma and the films that were additionally exposed to O and H atoms, respectively. Both O and H atoms reduce concentration of CHx and Si-$CH_3$ groups. Reduction of $CH_2$-$CH_2$ groups and formation of C-C=O

21

groups is more pronounced for oxygen plasma. As expected, the degree of surface hydrophilisation is smaller in the case of hydrogen plasma (see also [3,5]).

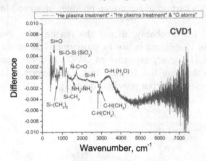 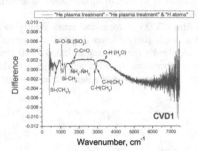

**Fig.6.** Differential FTIR spectra of two CVD1 samples exposed to He plasma (VUV radiation + ions), but one of the samples then was exposed to O atoms.
**Fig.7** Differential FTIR spectra of two CVD1 samples exposed to He plasma (VUV radiation + ions), but one of the samples then was exposed to H atoms.

## CONCLUSIONS

1. The surface densification of CVD1 material occurring in He plasma plays an important role in the reduction of plasma damage during the following exposure in oxygen plasma. However, the damage reduction is also observed in CVD2 material although the surface densification is not pronounced. Therefore in this case, the formation of surface recombination centers plays a key role. The nature of surface recombination centers will be analyzed and discussed in our future publications.

2. The formation of a dense layer is in good agreement with our previous data [3]. Surprisingly, we observed that this dense layer is mainly formed by hydrocarbons. The origin of the hydrocarbon compounds is not completely clear. One can assume that these hydrocarbons form as a result of modification of amorphous carbon like porogen residues observed in all CVD based SiCOH films [7].

3. The presented results also supported our previous conclusions that He/$H_2$ based resist strip processes are the most promising from point of damage free processing.

## ACKNOWLEDGMENTS

This study is carried out in the frame of MSU-IMEC Agreement and was supported by RFBR 09-02-01374 and 08-02-00465 and Key Science School (SS-133.2008.2).

## REFERENCES

1. M. R. Baklanov et al. Proc. of Int. Conf. Sol.-St. Integr. Circuit Techn. (ICSICT), p. 291, Shanghai, 2006.
2. T. V. Rakhimova et al. IEEE Trans. on Plasma Sci., Special Issue, TPS, 2009.
3. A. M. Urbanowicz et al., Electrochem. Sol. St. Lett., **10**, G76 (2007).
4. P. Lazzetti et al. Thin Solid Films, 516, 3697 (2008).
5. A. M. Urbanowicz et al. Electrochem. Sol. St. Lett., **12**, H292 (2009).
6. P. Marsik et al. Proc. of Advanced Metallization Conference in 2008, MRS, 2009.
7. P. Marsik et al. Proc. of Advanced Metallization Conference in 2008, MRS, 2009.

Mater. Res. Soc. Symp. Proc. Vol. 1156 © 2009 Materials Research Society

# Characterization of Plasma Damage in Low-k Films by TVS Measurements

Ivan Ciofi, Mikhail R. Baklanov, Giovanni Calbo, Zsolt Tőkei and Gerald P. Beyer
IMEC, Kapeldreef 75, Leuven, B-3001, Belgium

## ABSTRACT

Triangular Voltage Sweep (TVS) measurements were evaluated as a technique to characterize plasma damage in low-k films. For this study, we prepared blanket wafers with low-k films of different porosity and k-value. Before deposition, a thin layer of dry thermal oxide was grown on the n-type wafers to stabilize the silicon interface. After deposition, low-k films were exposed to $N_2/H_2$ plasma for different times in order to induce different degree of plasma damage. Untreated low-k films were included as a reference. For electrical measurements, metal dots were deposited on wafer pieces to fabricate Metal-Insulator-Semiconductor (MIS) planar capacitors.

TVS measurements were performed on the different samples at 190°C. On samples exposed to $N_2/H_2$ plasma, we detected a current peak in the TVS trace, whose magnitude increased with exposure time to plasma. No peaks were detected on untreated films. This indicates that TVS measurements are sensitive to plasma damage. TVS results correlated well with FTIR spectra that showed increasing damage and $H_2O$ uptake with increasing exposure time to $N_2/H_2$ plasma. We conclude that TVS measurements are suitable for characterizing the degree of plasma damage in low-k films and complete materials analysis, because with the help of TVS a link to leakage properties can be made. As an application, we used TVS measurements for evaluating restoration of $N_2/H_2$ plasma damaged low-k films by long $N_2$-bake at high temperature. FTIR and CV measurements were performed before and after treatments to evaluate changes in material structure and k-value, respectively. Our data show that long $N_2$-bakes at high temperature can partially restore leakage (TVS), k-value (CV) and hydrophobic properties (FTIR) of low-k films that are damaged by exposure to $N_2/H_2$ plasma.

## INTRODUCTION

Low dielectric constant (low-k) materials are being tested as line-to-line insulators in order to improve performance of advanced Cu damascene metallization. Low-k materials present a lower dielectric constant (k-value) than conventional silicon dioxide ($SiO_2$), which would enable to reduce line-to-line (parasitic) capacitance and realize faster interconnects. However, typical integration steps in the damascene process flow, such as plasma etch and strip, can significantly damage low-k materials. In these cases, original hydrophobicity is often lost. Therefore, low-k materials can adsorb water molecules during integration (e.g. cleaning, chemical-mechanical polishing, storage in clean room environment), which causes poor performance (high dielectric constant and leakage) and reliability (short lifetime) of final Cu/Low-k interconnects [1]. Plasma damage is commonly characterized on blanket wafers by using techniques that allow evaluating possible changes in material structure, such as Fourier Transform Infrared Spectroscopy (FTIR), Ellipsometry Porosimetry (EP), Contact Angle measurements [2]. Besides, Capacitance-Voltage (CV) measurements are performed to evaluate possible degradations of the k-value. However, for a complete electrical characterization of plasma damage, the impact on dielectric leakage must be also estimated.

In this work, TVS measurements on MIS planar capacitors were investigated as a technique to characterize plasma damage in terms of induced dielectric leakage. They are basically current-voltage measurements, obtained by applying a voltage sweep. For an ideal dielectric (leakage current is absent), the TVS trace consists of the only capacitor's displacement current which is flat, apart from the dip due to silicon depletion. For a real dielectric, the TVS trace is distorted by the leakage current, which adds to the capacitor's displacement current: the higher the leakage, the more the TVS trace is distorted. In this respect, TVS measurements can be used to characterize plasma damage in terms of leakage. In the following, we present TVS, FTIR and CV measurements on low-k films of different porosity and k-value that were damaged by exposure to $N_2/H_2$ plasma. Restoration of $N_2/H_2$ plasma damaged low-k films by long $N_2$-bakes at high temperature is also discussed.

## EXPERIMENTAL DETAILS

Table 1 summarizes the samples we prepared for our study. As low-k dielectrics we selected SiCOH materials of different porosity and k-value: an SiCOH material with 7% porosity and nominal k-value of 3.0 (Type A) to be integrated as intrametal dielectric in 200 mm wafer technology; two SiCOH materials from different suppliers with same porosity of 25% and same nominal k value of 2.5 (Type B and C) to be used in 300 mm wafer technology. Prior to deposition, a thin layer of Dry Thermal Oxide (ThOx) was grown on n-type silicon wafers to stabilize the silicon interface. ThOx thickness was 2 nm on 200 mm wafers and 5nm on 300 mm wafers. After deposition, low-k films were exposed to $N_2/H_2$ plasma to induce plasma damage. For sensitivity study, blanket wafers with 300 nm of film A were exposed to $N_2/H_2$ plasma for 5 sec, 20 sec and 35 sec, respectively, in order to induce different degree of damage. Besides, blanket wafers with same thickness (180 nm) of film A, B and C were exposed to $N_2/H_2$ plasma for the same time (45 sec) in order to investigate possible material dependency. Blanket wafers with untreated films were always prepared as a reference. For electrical measurements, wafers were diced in pieces of around 3 cm $\times$ 3 cm and metal dots were deposited on top of the low-k films in order to form MIS planar capacitors. A schematic cross-section of the test structures is shown in Fig. 1 together with the measurement configuration.

**Table I** Description of the samples

| Low-k film | Thickness | $N_2/H_2$ Plasma | Gate Material | Purpose |
|---|---|---|---|---|
| SiCOH, 3.0, 7%_Type A (on 200 mm wafers) | 300 nm | (Reference) | Ta | - TVS sensitivity to plasma damage |
| | | 5 sec | | |
| | | 20 sec | | |
| | | 35 sec | | |
| SiCOH, 2.5, 25%_Type B (on 300 mm wafers) | 180 nm | (Reference) | Ti | - Material dependency |
| | | 45 sec | | |
| | | (Reference) | | - Restoration by long $N_2$-bakes at high temperature |
| | | 45 sec | | |
| SiCOH, 2.5, 25%_Type C (on 300 mm wafers) | | (Reference) | | |
| | | 45 sec | | |

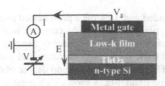

**Figure 1.** Schematic cross-section of the test structures and measurement configuration

TVS measurements were carried out at 190°C on a hot chuck probe station by an HP4140B pA meter/DC voltage source. Substrate to gate bias was swept from negative to positive voltages. Starting (negative) voltage for the different samples was set in order to have same initial electric field across the dielectric film of 1MV/cm. Finally, voltage sweep rate was 1V/sec and, unless otherwise specified, no bias stress was applied before initiating the voltage sweep. An HP4284A precision LCR meter was used for measuring (double sweep) CV curves at 25°C. FTIR measurements were performed in the range 400-5000 cm$^{-1}$ on blanket wafer pieces by using a BIORAD transmission spectrometer.

## RESULTS AND DISCUSSION

For an ideal dielectric, embedded in an MIS structure, the TVS current ($I_{TVS}$) consists of the capacitor's displacement current, given as the product of the low frequency capacitance ($C_{MIS}$) by the voltage sweep rate (dV/dt). Therefore, the TVS trace is flat, apart from the dip due to silicon depletion. For a real dielectric, dielectric leakage ($I_{Leakage}$) adds to the capacitor's displacement current and distorts the TVS trace:

$$I_{TVS} = C_{MIS}\frac{dV}{dt} + I_{Leakage} .$$ (1)

In (1), $I_{Leakage}$ accounts for all the possible leakage components, such as steady-state electronic current and ionic current. In the following, we report and discuss TVS, FTIR and CV measurements on the different samples. In the figures, TVS traces are normalized with respect to dot area and the sweep direction is indicated by the arrow. CV curves are normalized with respect to free-space capacitance, given as:

$$C_0 = \varepsilon_0 \frac{A}{T}$$ (2)

where $\varepsilon_0$ is the permittivity of free space (8.85 × 10$^{-12}$ F/m), A the dot area and T the low-k thickness, measured by ellipsometry. Normalized capacitance ($C/C_0$) directly provides the k-value when the MIS capacitors are in accumulation (positive gate bias).

### TVS sensitivity to N₂/H₂ plasma damage

Fig. 2 shows TVS measurements on film A for different plasma treatments. Blanket wafers with 300nm film were exposed to 5sec, 20 sec and 35 sec N₂/H₂ plasma, respectively. For a fare comparison, test structures were fabricated and measured together. Before initiating the voltage sweep, 1min bias stress at 1MV/cm was applied to the MIS capacitors (Ta dots). For the reference (untreated) sample TVS trace is flat, while it is distorted for samples treated with plasma. In particular, TVS trace is bended and presents a pronounced peak, whose magnitude increases with exposure time to $N_2/H_2$ plasma. TVS results correlate well with FTIR measurements on film A, performed to evaluate actual degree of damage for the different samples (Fig. 2). FTIR spectra indeed show that carbon depletion and water uptake increase with exposure time to plasma. Therefore, TVS peak magnitude and degree of damage as characterized by FTIR measurements follow the same trend. Fig. 3 shows TVS measurements on film A, B and C (Ti dots). Film thickness was 180nm and exposure time to $N_2/H_2$ plasma was 45 sec. As can be seen, similar (qualitatively) results are obtained for the three different low-k films, which validate the use of TVS measurements for characterizing plasma damage of low-k films. The different magnitude of the peaks in Fig. 3 indicates that degree of induced damage is different for the three different materials. As reported in previous works, TVS peak is probably due to protons ($H^+$), generated during the measurement itself by decomposition of water in the low-k film [3, 4, 5].

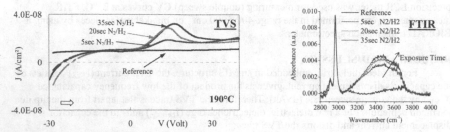

**Figure 2.** TVS and FTIR measurements on 300nm film A for 5sec, 20sec and 30sec $N_2/H_2$ plasma

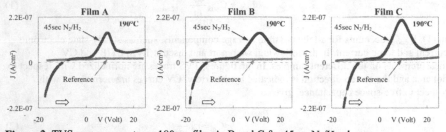

**Figure 3.** TVS measurements on 180nm film A, B and C for 45sec $N_2/H_2$ plasma

### Restoration of $N_2/H_2$ plasma damaged low-k films by long $N_2$-bake at high temperature

In order to investigate the possibility to restore $N_2/H_2$ plasma damaged low-k films by long $N_2$-bakes at high temperature, wafer pieces were baked at 350°C for 4h30min in $N_2$ atmosphere. A few pieces were measured immediately after baking. The remaining pieces were either left exposed to ambient for a few days or dipped in deionized $H_2O$ for a few hours to evaluate recovery of hydrophobic properties. The different treatments ($N_2$-bake, exposure to ambient, $H_2O$ dipping) were always performed on blanket wafer pieces. Metal (Ti) dots for electrical measurements were only deposited after the treatments. This study was performed for film A, B and C by using the samples that were damaged by 45sec $N_2/H_2$ plasma.

TVS, FTIR and CV measurements on the different samples are reported in Fig. 4 and 5. TVS traces are flat on baked samples. On samples that after baking were dipped in water for 3h, only a very small peak could be detected. Fig. 4 clearly shows that the magnitude of the TVS peak is significantly reduced after heat treatment and remains quite stable even after $H_2O$ dipping. This indicates that for baked samples leakage properties are partially recovered. In addition, FTIR spectra show that water uptake is significantly reduced after bake and that initial level could not be reestablished after water dipping. This indicates that material hydrophobicity could be partially restored. Finally, CV measurements performed on baked pieces after 6 days of exposure to ambient showed a reduced k-value (Fig. 5). Consistently, FTIR spectra showed a significant reduction of $H_2O$ content soon after baking and only minor reincorporation of $H_2O$ after 6 day of exposure to ambient (Fig. 5). Therefore, long $N_2$-bake at high temperature can partially restore leakage (TVS), k-value (CV) and hydrophobic properties (FTIR) of damaged low-k films. It is worth pointing out that shorter annealing times, such as 1h30min at 350°C, were not sufficient to recover leakage properties for all the three materials. In particular, quite pronounced TVS peaks could be still detected for film B and C of higher porosity. We conclude that typical annealing steps in the Cu damascene process flows (300°C – 400°C for 30sec – 3 min) might be too short to be effective for low-k materials with high porosity. As far as restoration mechanism is concerned, the long annealing at high temperature induces condensation reactions in damaged low-k materials, such as:

$$SiOH - SiOH \rightarrow Si - O - Si + H_2O \qquad (3)$$

Siloxane groups are intrinsically hydrophobic and quite stable with exposure to ambient of increased humidity [6]. Since plasma damage is expected to be more significant for highly porous low-k materials, long thermal annealing might be an effective solution for meeting the strict performance requirements for next generation interconnects.

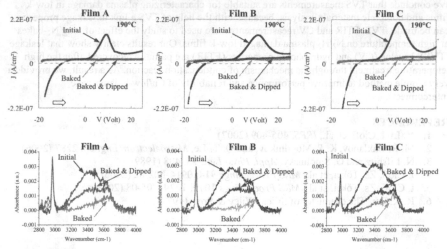

**Figure 4.** Restoration by a long $N_2$-bake at 350°C for 4h30min: TVS and FTIR measurements

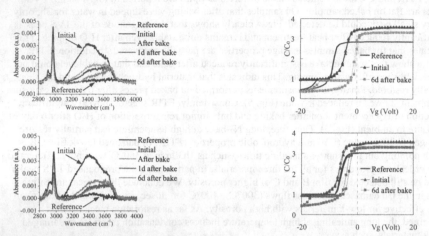

**Figure 5.** Restoration by a long $N_2$-bake at 350°C for 4h30min: FTIR and CV measurements

## CONCLUSIONS

TVS measurements on MIS planar capacitors were investigated as a technique for characterizing plasma damage in low-k films. We show that TVS measurements are sensitive to plasma damage. In particular, on samples exposed to $N_2/H_2$ plasma we detected a current peak, whose magnitude increased with degree of damage that was quantified by FTIR measurements. We conclude that TVS measurements are suitable for characterizing plasma damage in low-k films and complete materials analysis, because with the help of TVS a link to leakage properties can be made. TVS, FTIR and CV measurements were used to study the effect of long $N_2$-bakes at high temperature on $N_2/H_2$ plasma damaged low-k films. Our results clearly show that leakage (TVS), k-value (CV) and hydrophobic properties (FTIR) are partially recovered after the high temperature treatment through the mechanism of condensation reaction. As a result, we provide a very simple method to improve performance and reliability of Cu/low-k damascene interconnects.

## REFERENCES

1. Y. Li, I. Ciofi, et. al., *IRPS*, 405-409 (2007)
2. M. R. Baklanov, K. P. Mogilnikov, and Q. T. Le, *Microelectron. Eng.* **83**, 2287 (2006)
3. N. Lifshitz and G. Smolinsky, *Appl. Phys. Lett.* **55**, 408 (1989)
4. I. Ciofi, Zs. Tőkei, et al., *MRS Proceedings* **914**, 0914-F02-02 (2006)
5. I. Ciofi, Zs. Tőkei, et al., *MRS Proceedings* **1079**, 1079-N05-08 (2008)
6. R. K. Iler, The chemistry of Silica, Wiley & Sons (1979)

# Low-k Dielectrics II

Mater. Res. Soc. Symp. Proc. Vol. 1156 © 2009 Materials Research Society          1156-D02-05

# Effects of Polymer Material Variations on High Frequency Dielectric Properties

Gregory T. Pawlikowski
Tyco Electronics Corporation, P.O. Box 3608, Harrisburg, PA 17105

## ABSTRACT

In high frequency signal packaging, the plastic dielectric material takes on an increasingly important role in the performance of the signal transmission. Variations within the plastic can occur as a result of a number of manufacturing and environmental processes. These variations can be sufficient to change the dielectric properties. The extent of change to the dielectric properties of polymer materials as a result of controlled variations to major filler additives, moisture and temperature has been investigated. The combined effects of moisture and temperature can cause changes to the dielectric constant of certain materials by more than 30 %.

## INTRODUCTION

Traditionally, the electrical performance assessment of plastic materials used in electronic packaging focused simply on their insulative nature (high dielectric strength) and their ability to physically isolate adjacent metal conductors. As the frequency of the electrical signals in circuits increases and the spacing between conductors decreases, the electromagnetic properties in the plastic materials play an increasing role in the propagation of the electrical signal.

Two important electromagnetic properties in plastic materials are the relative permittivity (or dielectric constant) and dissipation factor (or dielectric loss tangent). The appropriate dielectric constant of a polymer along with the correct geometric design of the system helps reduce the signal-propagation delay and distortion through the electronic package, and minimize cross-talk. Materials with a low dielectric loss tangent can reduce attenuation in circuits, and can improve signal integrity.

When designing high speed, high frequency connectors, it is critical to couple the dielectric properties of the plastic housing material with the geometric design in order to obtain the appropriate electrical performance over the desired frequency range. The potential flaw in this design approach is the assumption that the dielectric properties of a specific plastic resin grade will remain constant. A specific grade of plastic resin can undergo a number of changes that can cause shifts in the dielectric constant (Dk) and loss tangent (tan δ) of the material. These dielectric properties of plastic materials are directly influenced by the chemical make-up of the resin formulation (that can include the base polymer, reinforcing fillers, flame retardants), and to a lesser extent processing modifiers, colorants, plasticizers, lubricants, etc. Lot-to-lot variations in the concentrations of the major resin additives (glass fillers and flame retardants) can cause changes in the dielectric properties of a particular resin grade.

The dielectric properties of plastic resins are also influenced by the orientation, density and mobility of the dipole moments of polymer repeat units. Even resins from within a specific lot of material can be processed in ways that can lead to variations in polymer and reinforcing fiber orientation, degree of crystallinity, and density. Such variations can cause changes in the Dk and tan δ as a result of different molding parameters. Even after a product has been manufactured, there are environmental factors that can impart changes in a resin that can affect the dielectric properties. Increases in temperature can increase the mobility of polymer chains, especially thermal excursions through a plastic's glass transition temperature. As the largest contribution to the polarizability of polymers, the orientational polarizability results from a change in alignment of the permanent dipole moments in the polymer to that of an applied electric field by the physical movement of the group associated with the dipole [1]. Thus the increased mobility of the polymer chains at elevated temperatures can lead to changes to Dk and tan $\delta$.

Moisture absorption in a resin can have a combination of effects on the dielectric properties. Moisture can act as an additional component in the material that can have an additive effect on the dielectric properties. With the Dk of water being ~ 50 at a frequency of 10 GHz [2], a relatively small absorption of moisture into a resin can cause a significant change. In materials such as polyamides that have an affinity for moisture, water can induce plasticization in the resin. This can cause a shift in the glass transition temperature of the material to lower temperatures, increasing the mobility of the polymer chains even at room temperature. Again, a change in the mobility of the polymer chains can cause a change in the dielectric properties.

There is evidence in the literature that at least some of the factors mentioned above can significantly change the Dk of materials [3] [4] [5]. To what extent these variations affect the dielectric properties of thermoplastic materials is not well characterized, especially at high frequencies (> 1 GHz). If the dielectric properties of the plastic material change, the electrical performance of a connector may suffer. This paper seeks to characterize the changes to Dk and tan $\delta$ associated with variations in materials caused by moisture absorption, reinforcing filler quantity, flame retardant content, and temperature.

## EXPERIMENTAL

Evaluations of the effects of glass filler and flame retardant content on the dielectric properties used polyamide 4-6 type material supplied by DSM Engineering Plastics. The grades studied were Stanyl® TE300 (unfilled), Stanyl® TE200 F6 (with a 30 wt.% glass fiber content), and Stanyl® TE250 F6 (with 30 wt.% glass fiber and an undisclosed flame retardant). Using these three resin grades allowed for the isolation of the material variations related to the major additives. And in some cases, mixtures of these grades provided intermediate levels of the additives. The materials used to study the effects of moisture and temperature were two polyamides (Stanyl® TE200 F6 from DSM Engineering Plastics and Zytel® HTN FR52G30 from DuPont®), three polyesters (Valox® 420 SE0 from SABIC Innovative Plastics, and Thermx® CG933 PCT & Zenite® 6130L from DuPont®), a syndiotactic polystyrene (Questra® EA535 from Dow Chemical), and 3112 RTV silicone from Dow Corning®.

Two techniques were used for measuring the dielectric properties of the plastic materials. A stripline method was used when studying the effects of moisture [6]. Using this technique, Dk and tan $\delta$ data was obtained over a frequency range between 0.1 to 20 GHz. Samples analyzed using the stripline technique consisted of a length of copper conductor sandwiched between strips of the plastic, and were constructed by a custom process. This method provided a single monolithic structure with the center conductor strip tightly surrounded by the dielectric material. Samples approximately 12 mm wide and 2 mm thick of various lengths were utilized for the testing.

A second method for measuring Dk and tan $\delta$ in this study was through a split-cylinder resonant cavity technique. In this technique, a thin sample is placed in the gap between the two shorted cylindrical waveguide sections. Resonant measurement methods provide an accurate means of obtaining Dk and tan $\delta$, but they typically provide data at only one frequency.

However, unlike the stripline method, the split cavity resonator is more easily adapted for higher temperature measurements. Disks measuring 30 mm in diameter and approximately 1 mm thick were used when analyzing the dielectric properties using the resonant cavity technique. These samples were injection molded using consistent molding parameters.

In varying the moisture content, ovens were used to accelerate either the drying or moisturization of the samples. A set of five samples of each material were dried in a forced air convection oven for 3 days at 90 °C. A separate set of five samples were moisturized to the point of saturation using a humidity chamber set at 85 °C/85 % R.H. for a duration of 7 days. The dried and moisturized samples were stored in moisture barrier bags between testing. Moisture content in the dry samples was measured by a Computrac® VaporPro® moisture analyzer from Arizona Instruments. The moisture content in the ambient and moisturized samples was determined by gravimetric methods relative to the dry mass of each sample.

## RESULTS AND DISCUSSION

The stripline method was used to measure the effect of moisture absorbed into a plastic resin. Moisture was found to have a significant effect on the dielectric properties of the polymers in this study. The Stanyl® polyamide, Thermx® and RTV silicone each showed an increase in their dielectric constants and tan δ across a broad spectrum of frequencies. An example of this behavior is shown in the plots of Dk and tan δ vs. frequency for the Stanyl® resin that was tested when dry, at ambient storage conditions and when moisturized (Figure 1). The ambient measurement was not made on the silicone material. While the magnitude of the effect of the added moisture on Dk decreases with increasing frequency, it remains significant in the 10 to 20 GHz range.

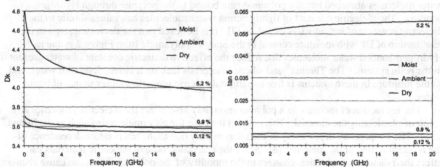

**Figure 1.** Shifts in the broadband Dk and tan δ for the Stanyl® polyamide material (moisture contents listed as wt.%).

The rate of the change in the dielectric properties with increasing moisture content at 10 GHz for the Stanyl®, Thermx® and silicone using the stripline method is shown in Figure 2. The increase in the dielectric constant with increasing moisture content fits a linear relationship reasonably well for each of the resin types. At 10 GHz, the Dk of the Stanyl®, Thermx® and silicone increased by 15 %, 12 % and 8 %, respectively. Much larger increases were observed for the effects of moisture on tan δ (575 %, 260 % and 40 %, respectively). The change in Dk and tan δ was found to be fully reversible. When the moisturized Stanyl® sample was re-dried to

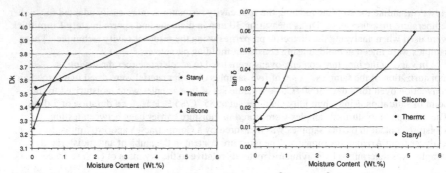

**Figure 2.** Effect of moisture on the Dk and tan δ for Stanyl®, Thermx®, and silicone resins at 10 GHz.

remove moisture, the Dk and tan δ returned to their pre-moisturized levels.

The impact that the moisture has on the dielectric properties varies for the different polymer materials, expecially the Stanyl® polyamide. This may be due in part to the interaction water has with polyamides compared to the other materials. The impact of moisture on the Thermx® and RTV silicone materials is consistant with what would be expected of free (unbound) liquid water, if the effect of the moisture was strictly additive (based on a Dk of 100 % water equal to 50). The Stanyl® polyamide material deviates from this correlation with a lower rate of Dk increase with increasing moisture content. This is likely due to the fact that some fraction of the water molecules absorbed into the polyamide are bound to the polymer through Hydrogen-bonding. The dielectric constant of tightly bound water molecules has values similar to the dielectric constant of ice, which is only 3.16 [7]. It is believed that the decreased slope in the relationship of Dk with moisture content in the Stanyl® polyamide from Figure 2 is due to Hydrogen-bonded water molecules that reduce the effective dielectric constant of water present in the resin system. The Thermx® and silicone materials lack such interactions with water, and so the moisture in these systems is free to exhibit dielectric properties closer to that of liquid water.

The interaction of moisture in a polymer resin may also affect the dielectric properties as a result of plasticizing the material. This is again the case for polyamide resins such as the Stanyl® resin. The temperature dependency of the Dk is mainly due to the temperature dependence of the relaxation time of the materials [8]. Since the relaxation time of dielectric materials can effect their dielectric properties, changes to the mobility of the polymer chains can cause changes in the Dk and tan δ.

The consequence of this is that the impact of moisture that was observed for the Stanyl® polyamide in Figure 2 may be a combination of effects from 1) the additive contribution of the water molecules, and 2) the increased mobility of the polymer chains as a result of water's plasticizing effect. In order to attempt to separate out the effects that moisture has on the dielectric properties of polymer systems, the Dk of a number of polymer resins were measured at elevated temperatures both in the dry and moisturized states. Using a split-cylinder resonator that was fitted with a heated jacket, the Dk of the materials were measured at temperatures between 25 and 125 °C and at a fixed frequency of 16 to 17 GHz (depending on the polymer being measured).

The changes to the Dk of the various materials with respect to temperature are plotted in Figure 3. While there is a small increase in the Dk with increasing temperature for the Zytel® and Stanyl® polyamide materials, the Dk of the other materials remain largely unaffected by even a 100 °C increase in temperature. This, despite the fact that each of the materials exhibits a glass transition temperature ($T_g$) from a glassy to a rubber, more mobile state within the temperature range tested. It may have been expected that the Dk of the materials would have seen a greater increase when the test temperatures exceeded their glass transition temperatures. The reason for lack of increase in Dk may lie in the fact that, as in mechanical testing, that apparent $T_g$ of a material is dependent on the frequency at which the material is tested.

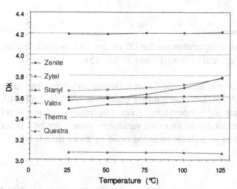

**Figure 3**. Effect of temperature on the Dk of various dry materials at ~ 16.5 GHz.

Figure 4 shows the effects of the Dk of dry and moisturized materials at various temperatures. With the exception of the polyamide materials, the Dk of the moisturized materials were unaffected by temperature. There was a small shift in the Dk due to the additive effect of moisture, but like the dry material, the Dk did not increase with temperature. In the case of the polyamide resins (Zytel® and Stanyl®), there was a significant increase in the impact that moisture has on Dk at higher temperatures. Whereas the increase in Dk for the Zytel® and Stanyl® polyamides due to moisture absorption was 7 % and 12 %, respectively at 25 °C, the increase in Dk was much larger (12 % and 28 %, respectively) when measured at 125 °C. The polyamide materials showed a steady, linear increase in their response rates of Dk with temperature. Since the change in Dk at the various temperatures is a direct result of the increased moisture in the materials, it is believed that the increasing impact that temperature has on Dk at elevated temperatures is a function of the dielectric behavior of water.

The effects that major filler materials have on the dielectric properties of a polymer resin were also studied using the resonant cavity method. Specifically, the change in the Dk and tan δ as a function of glass fiber and flame retardant contents were measured on the Stanyl® resin at ambient conditions. The Dk exhibited a moderate increase with increasing glass fiber content, from 3.27 (unfilled) to 3.67 (30 wt. % glass fibers). The tan δ remained relatively unchanged compared to the effects of moisture and temperature, decreasing from 0.0151 to 0.0133. Taking into account that the lot to lot variation of reinforcing glass fiber should not vary by more that ± 2 or 3 wt.%, the effect that variations in glass fiber content might have on the Dk and tan δ of glass fiber reinforced resins is less than 1 %.

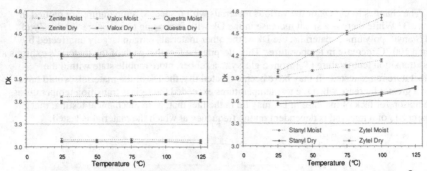

**Figure 4**. The effects of temperature on the Dk of dry and moisturized samples of Zenite®, Valox®, and Questra® (a); and of Stanyl® and Zytel® (b).

As with the glass fiber content, the flame retardant exhibits a moderate increase in dielectric constant (from 3.62 to 3.93), but only a minor change in tan δ (from 0.0133 to 0.0141) compared to the effects of moisture and temperature. Even a nominal variation in relative flame retardant content of 5 weight percent due to lot to lot variations would not be expected to cause a change in the dielectric properties of a flame retarded resin by more than 0.5 %.

## CONCLUSIONS

While typical variations in major additives (such as reinforcing glass fibers and flame retardants) in polymer resins have little effect on the dielectric properties of the materials, variations due to environmental factors such as moisture and temperature can have significant effects. The impact of moisture on Dk and tan δ is significant for all polymer materials tested. This impact of the moisture is diminished in materials that absorb and interact with moisture such as polyamides, but the reduced effect is offset by the relatively large amount of moisture that they can absorb. Polymers that don't associate with moisture may show a greater effect on the increase in Dk, but this is mitigated by the small amount of moisture that may be absorbed into these materials.

The dielectric properties of polymers like polyesters and polystyrene, that do not significantly absorb moisture, are surprisingly stable over a broad temperature range up to 125 °C. It is believed that the delay in the materials' response to the high frequency measurements shifts the apparent glass transition temperatures to higher values, and prevents a more significant increase in the dielectric properties when the materials are tested above their accepted $T_g$ values. When moisture is present in the system, an increase in temperature can lead to a dramatic increase in the dielectric properties. In materials that absorb a significant amount of moisture, the combined effects of moisture and increased temperature may have a synergistic action that causes a greater increase in the Dk than the material would experience with either moisture or temperature as separate factors.

# REFERENCES

1. F. Mercer and T. Goodman, High Perform. Polym., **3**(4), pp. 297-310 (1991).
2. E. Nyfors and P Vainikainen, *Industrial Microwave Sensors*. Norwood, MA, Artech House, 1989, pp. 41-49.
3. J.-M. Heinola and J.-P. Ström, Elect. Insul. Mag., **23**(3), p. 23-29, (2007).
4. C. Zou, J. Fothergill and S. Rowe, IEEE Transact. Dielect. Elect. Insul., **15**(1), pp.106-117, (2008).
5. S. Singha and M. J. Thomas, IEEE Transact. Dielect. Elect. Insul., **15**(1), pp. 12-23 (2008).
6. C. Morgan, DesignCon08, 5-TA4, Santa Clara, CA. 2008.
7. V. F. Petrenko and R. W. Whitworth, *Physics of Ice*, Oxford, Oxford University Press, 2005, p. 95.
8. K. C. Kao, *Dielectric Phenomena in Solids*, London, Elsevier Academic Press, 2004, pp. 97-99.

Mater. Res. Soc. Symp. Proc. Vol. 1156 © 2009 Materials Research Society        1156-D02-08

# Optimization of Low-k UV Curing: Effect of Wave Length on Critical Properties of the Dielectric

German Aksenov[1], D. De Roest[2], P. Verdonck[1], F. N. Dultsev[3], P. Marsik[4], D. Shamiryan[1], H. Arai[5], N. Takamure[5], M. R. Baklanov[1]

[1] IMEC, Leuven, Belgium
[2] ASM Belgium, Leuven, Belgium
[3] Institute of Semiconductor Physics, Novosibirsk, Russian Federation
[4] Masaryk University, Brno, Czech Republic
[5] ASMJ, Tokyo, Japan

## ABSTRACT

The results of recent investigations show that after UV curing of CVD SiCOH low-k films deposited with organic material (porogen) some amount of the porogen remains in the cured films in the form of non-volatile graphitized phase, known as "porogen residue". These residues could influence leakage current and reliability. The goal of the present work is investigation of the different parameters of UV curing that can influence amount of the porogen residue. In this work we focused generally on the study of the amount of porogen residues as function of the wavelength of curing light and the porosity of the material (amount of deposited porogen). To study the curing dependence on the wavelength, we compared optical properties (measured by spectroscopic ellipsometry) and IR adsorption (measured by FTIR) of samples cured by 172 nm monochromatic light (lamp A) with samples cured by broadband source with wavelength more than 200 nm (lamp B). To understand how the amount of porogen residue depends on the amount of deposited porogen (porosity), three films with different k-value were deposited: a film with k = 3 deposited without porogen and two porogen-based low-k with target k-value of 2.5 and 2.3. Furthermore, taking into account that He/H2 plasma effectively removes the porogen residues from porous films without any plasma damage of the matrix material, we exposed the films to that plasma. Then these films were cured by broadband lamp at different temperatures and amount of porogen residues was measured by ellipsometry. It was found that He/H2 plasma cannot fully remove the porogen and causes film shrinkage. The Subsequent UV curing does not produce significant changes.

## 1. INTRODUCTION

UV light combined with thermal annealing is widely used for porogen removal and improvement of mechanical properties of porogen-based low-k materials. However, some issues are still not sufficiently studied and understood. For instance, UV assisted structural changes in the deposited porogen material are associated with the effusion of light hydrogen-rich components and thermal graphitization. Non-volatile graphitized phase remains inside the low-k film and is referred to as "porogen residue". These residues can induce a current leakage in low-k films. Quantitative study of the porogen residues (amount, properties and composition) is challenging because of limited sensitivity of FTIR spectroscopy to such compounds. However, precise analysis shows that this residues has IR absorption near 1600 cm-1, which is typically attributed to conjugated C=C bonds (for instance in the classical conducting polymer (-C=C-

C=C-C=C-). Therefore, the problems related to leakage and reliability are quite obvious (see figure 4).

In this work, we are reporting the results of the analysis of low-k films cured with UV light at different conditions. The major part of this analysis is based on the data obtained by UV ellipsometry that allows the analysis of absorption spectra of low-k films. UV ellipsometry is also very sensitive to the porogen residues. By using low-k samples with different time of curing, we show that the porogen residue significantly increases the leakage current of the low-k material. It is shown that UV spectroscopy is an efficient way to study formation and behavior of porogen residues.

Recently it was reported [6], that He/H2 plasma effectively removes the porogen residues from porous films without any plasma damage of the matrix material. In order to check the possibility to remove all porogen by He/H2 plasma only we exposed the films to that plasma. In addition, these films were cured by broadband lamp at different temperatures. After every step (deposition, plasma treatment, curing) the amount of porogen residues was measured by spectroscopic UV ellipsometry.

## 2. EXPERIMENTAL DETAILS.

SiCOH porous low-k films with different k value deposited by PECVD, was used to study the effect of UV curing. The films with k = 3 were deposited without porogen (just SiCOH matrix). The films with k = 2.3 and 2.5 were deposited mixing SiCOH matrix precursor with organic porogen precursor.

Then a part of samples with k = 2.3 were cured for various curing times: 0.2T, 0.5T, T, 2T and 5T, where T is the optimum time (corresponds to the minimum of k-value), by monochromatic light with 172 nm wavelength and by a broadband lamp with wavelength above 200 nm. The goal of this experiment was to observe the effects of the UV curing on optical and structural properties of the low-k films. Nitrogen-purged UV ellipsometry in a range from 150 to 750 nm was used to observe the changes of the dielectric function of the cured material while FTIR in a nitrogen atmosphere was used to observe structural changes of the cured material. To measure the leakage current as a function of different curing time and electrical field, the same films were deposited on high-doped wafers and cured at the same conditions. The leakage current was measured by a mercury probe.

The next set of samples with k = 3, k = 2.5, k = 2.3 was cured by the lamp B to evaluate the amount of porogen residues as function of deposited porogen.

For experiment with plasma treatment and subsequent curing at different temperatures 60 nm thick films were used with target k-value of 2.3. The conditions of plasma treatment were the following: 18% H2, 82% He, 530 Pa, 900 W, at 370 C for 30 seconds. This was the "traditional" He/H2 plasma that was already widely studied before. Then four wafers which were exposed to He/H2 plasma and five wafers without He-H2 plasma treatment received a UV cure by the lamp B at different temperatures.

All ellipsometric measurements were analyzed using Lorentz oscillator model.

## 3. RESULTS AND DISCUSSION

The dispersions of optical properties of the low-k films cured with the lamp B are shown in figure 1. The time evolution of the spectra can be separated into two stages with characteristic behavior. In the early stage of the UV-curing process, the rapid removal of significant volume of the porogen takes place, manifesting itself as a significant decrease of the extinction, particularly at 200 nm. This band was previously attributed to porogen [1, 2], due to correlation with the removal of the adsorption peak of CHx groups around 2900 cm-1 from the FTIR spectra (see figure 2b). For longer curing times, UV light causes matrix shrinkage and reduction of porosity. An increase of the extinction at 270 nm can be related to porogen residues formed in the films during the UV curing.

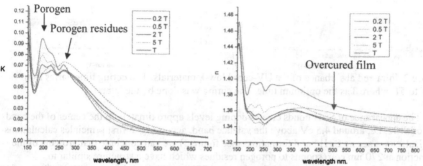

Figure 1. Optical functions of the UV-cured low-k materials. The curing time varies from 0.2T to 5T, where T is the optimum time. Curing was done by lamp B.

The figures 2 and 3 show FTIR spectra of the low-k films cured with the lamp B for different times. Si-H absorption peak does not appear during the curing with wavelength more than 200 nm. There is no decrease of the Si-CH3 peak. It was found that UV light with wavelengths shorter than 190 nm (lamp A) efficiently improves mechanical properties of the low-k films but breaks Si-CH3 bonds, forms Si-H groups, and creates significant amount of amorphous-carbon-like porogen residue [1, 2, 3]. In comparison with the films cured by the lamp A, UV-light with wavelengths longer than 200 nm (lamp B) provides similar improvement of mechanical properties without any reduction of Si-CH3 (see figure 2a) and without formation of SiH groups (see figure 3). The amount of porogen residue after the lamp B cure is much lower than that after the lamp A cure, although it increases with overcuring time. These findings are supported by the results of quantum-chemical calculations [3] that showed the existence of a threshold of photochemical reactions located near 190-200 nm. FTIR data and Molecular Mechanics simulations allowed us to conclude that the photochemistry of the curing done by the lamp B is principally different from the lamp A and occurs without reduction of concentration of (CH3)3-SiO1/2 (M) and (CH3)2-SiO2/2 (D) groups.

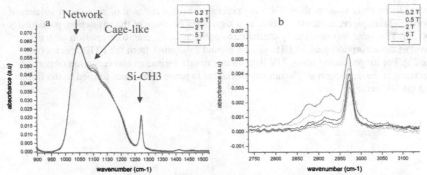

Figure 2. Infra-red absorbance of the UV-cured low-k materials. The curing time varies from 0.2T to 5T, where T is the optimum time. The curing was done by the lamp B.

The appearance of Si-H bonds creates doping levels approximately in the center of the band gap of SiO2, i.e. around 4.5 eV above the valence band, according to first principles calculations [4]. But, as we see no Si-H bond formations (see figure 3) we can conclude that growing of the extinction at 270 nm corresponds to porogen residues which have a structure similar to amorphous carbon, with characteristic band around 270 nm originated from π-π* transitions of sp2 bonded carbons [1, 2, 5].

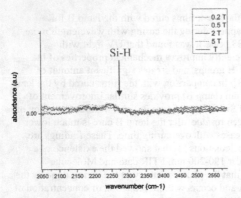

Figure 3. Infra-red absorbance of the UV-cured low-k materials. The curing was done by lamp B, curing time varies from 0.2T to 5T, where T is the optimum time.

Figure 4 shows a leakage current in low-k films as a function of curing time and electrical field (a), and the IR absorbance of the porogen cured at different times (b). Increase of the C=C peak during the UV-curing correlates with the increase of the UV-extinction at 270 nm. It can be concluded that the increase of porogen residue cause an increase in the leakage current.

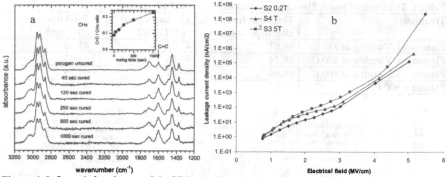

Figure 4. Infra-red absorbance of the UV-cured porogen (a) and the leakage current of the low-k films cured for different times by the lamp B (b).

The amount of porogen residues depends on the amount of deposited porogen. Figure 5 shows optical properties of the films deposited with different amount of porogen (the amount for the film with k = 2.3 is higher than for the film with k = 2.5). The film with k = 3.0 were deposited without any porogen. One can see that the amount of porogen residues strongly increases for the films deposited with higher amount of the porogen precursor, and that the film deposited without porogen does not have any porogen residues.

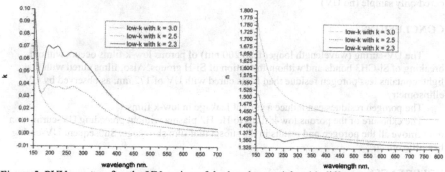

Figure 5. PUV spectra after the UV-curing of the low-k materials with different target k-values.

Table I shows the thickness of the low-k films which were treated by He/H2 and than cured by lamp B at different temperatures, and one film which was thermally cured only. The optical properties of those films are shown in figure 6. One can see that the plasma treatment causes substantial film shrinkage. Refractive index of the treated films decreases quite a lot after exposure to He/H2 plasma due to porogen removal (see figure 6). But some amount of the porogen and the porogen residue remain in the film. Subsequent UV-curing does not reduce the remained porogen.

43

Table I. Thickness of low-k films after plasma treatment and UV-curing.

| Treatment | Thickness (nm) after deposition. | Thickness (nm) after He-H2 | Thickness (nm) after cure |
|---|---|---|---|
| Plasma + UV curing at 250 C | 66.61 | 47.05 | 45.98 |
| Plasma + UV curing at 400 C | 66.54 | 46.65 | 45.05 |
| Plasma + UV curing at 500 C | 66.64 | 46.87 | 44.78 |
| Plasma + 500 C | 66.63 | 46.97 | 45.59 |

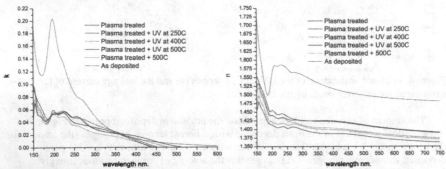

Figure 6. Optical properties of the plasma treated and the UV-cured low-k materials. The curing temperature varies from 250C to 500C. One curve (dotted line) corresponds to the thermally cured only sample (no UV).

## CONCLUSIONS

The UV-curing (wavelength longer than 200 nm) of porous low-k films occurs without breaking of Si-CH3 bonds and without formation of Si-H groups. Also, films cured with that light contains less porogen residue than films cured with UV of 172 nm, as observed by UV ellipsometry.

The porogen residues can induce a current leakage in low-k films.

The exposure of the porous low-k films to He/H2 plasma without preceding UV-curing can not remove all the porogen and results in the substantial film shrinkage. Subsequent UV-curing has little effect.

## REFERENCES

1. P. Marsik, et al., PSSc, 5, 1253-1256, (2008).
2. S. Eslava, et al., Journal of the Electrochemical Society 155, G115-G120 (2008).
3. L. Prager et al., Microelectronic Engineering 85, 2094–2097 (2008).
4. C. G. Van de Walle, Physica B 376(1-6), (2006).
5. S. Kassavetis, et al., Diamond and Related Materials 16, 1813-1822 (2007).
6. A.Urbanowicz, et al., Proc. Of Advanced Metallization Conference, 20-22 Sep. 2009, San Diego, USA

Mater. Res. Soc. Symp. Proc. Vol. 1156 © 2009 Materials Research Society          1156-D02-09

# Application of UV Irradiation in Removal of Post-Etch 193 nm Photoresist

Q. T. Le[1], E. Kesters[1], L. Prager[2], M. Lux[1], P. Marsik[1], and G. Vereecke[1]

[1] Imec, Kapeldreef 75, 3001 Leuven, Belgium

[2] Leibniz-Institut für Oberflächenmodifizierung, Permoserstr. 15, 04318 Leipzig, Germany

## ABSTRACT

This study focused on the effect of UV irradiation on modification of polymethyl methacrylate-based photoresist, and then on wet photoresist (PR) removal of patterned structure (single damascene structure). Three single-wavelength UV sources were considered for PR treatment, with $\lambda = 172$, 222, and 283 nm. Modification of blanket PR was characterized using Fourier-transform infrared spectroscopy (FTIR; chemical change), spectroscopic ellipsometry (SE; thickness change), and dissolution in organic solvent (solubility change). While for patterned samples, scanning electron microscopy (SEM) was used for evaluation of cleaning efficiency. In comparison to 172 nm, the PR film irradiated by 222 nm and 283 nm photons resulted in formation of higher concentration in C=C bond. Immersion tests using pure $N$-methyl pyrrolidone (NMP) at 60 °C for 2 min showed that some improvement in PR removal was only observed for PR films treated by 283 nm UV for short irradiation times. Irradiation by photons at the other two wavelengths did not result in an enhancement of removal efficiency.

The PR film treated by 222 nm photons was chosen for further study with $O_3/H_2O$ vapor at 90°C. Experimental results showed a complete PR and BARC removal for UV-treated PR, which can be explained by C=C bond cleavage by the oxidizer.

## INTRODUCTION

In Back-End-of-Line processing, the remaining photoresist (PR) layer after plasma etch is traditionally removed using plasma process. However, dry ashing of PR degrades porous low-$k$ dielectrics [1-3]. To minimize damage to low-$k$ material, wet alternative methods using solvent for removal of PR layer are gaining a renewed interest. The etch plasma used to pattern the dielectric stack leads to formation of a "crust" layer at the PR surface. The presence of the crust makes a complete removal of PR impossible using a pure organic solvent, especially for small features, i.e., ½ pitch ≤90 nm. Indeed, the crust, most likely composed of cross-linked polymer, is not soluble in organic solvents [4-6]. PR removal becomes even more challenging with reduced critical dimensions. A wet process combining the chemical action of organic solvents (bulk PR dissolution) with physical forces (for mechanically removal of plasma-induced crust) was shown to be able to remove PR on a structure of 150 nm ½ pitch but not for 90 nm ½ pitch [5,7].

Over the past years, ozone dissolved in water was shown to be an efficient process in the removal of organic contamination and PR layers [8, 9]. This process is regarded as a more environmentally friendly process, with lower costs, compared to traditional processes.

In a recent study [10], we have shown that UV treatment of PR at certain wavelengths generates C=C bonds, and represents a convenient way to increase the concentration in C=C bonds that are potential sites for chemical reaction with an oxidizer. Therefore, UV treatment of PR can be used as an intermediate step to enhance PR removal using oxidizing chemistries.

This study investigates the modification of plasma-treated (post-etch) PR film by UV irradiation and removal of plasma-treated PR and BARC by exposure to $O_3/H_2O$ vapor followed by a rinse using organic solvent. A comparison will be made between the case without and with UV treatment.

## EXPERIMENT DETAILS

The PR under study was a commercial DUV (193 nm) poly methyl acrylate/methacrylate -based resin, with lactone and adamantane groups in the side chains to enhance certain chemical and physical properties of the PR film. Typically, a photoresist (~150 nm)/BARC (~33 nm) layer was coated onto a single damascene structure consisting of TiN hard mask/low-$k$ dielectric/SiCN/Si stack (90 nm ½ pitch). For both blanket PR and damascene structures, a three-step plasma etch was performed (aimed at BARC and hard mask opening) at room temperature prior to UV treatment and wet clean, including (a) HBr/60 s, (b) $Cl_2/O_2/20$ s, and (c) $Cl_2/HBr/16$ s, using a Lam Research dual-frequency dielectric etch chamber. The low-$k$ dielectric used in this study was a CVD porous SiOCH material with a target $k$-value of 2.5 and porosity ~25 %.

UV radiations were generated using different sources of three wavelengths: a $Xe_2^*$ excimer lamp ($\lambda = 172 \pm 12$ nm), a narrow-band KrCl excimer lamp ($\lambda = 222 \pm 1.2$ nm), or a XeBr* excimer lamp ($\lambda = 283$ nm), all supplied by Heraeus Noblelight GmbH. The UV treatments were done under vacuum. The effect of UV irradiation in terms of PR removal was assessed in two ways: (a) direct PR removal of blanket plasma-treated PR carried out in a beaker set-up for a constant time of 2 min using pure $N$-methyl pyrrolidone (NMP) at 60°C, and (b) reaction with an oxidizer, i.e. by exposure of 222 nm UV-treated PR to $O_3/H_2O$ vapor at 90°C followed by a rinse in either propylene carbonate (PC) at 90°C [11] or in de-ionized water for 1 min. Both blanket plasma-treated PR and damascene structure were used for case (b). These experiments were done in a home-made, small reactor consisting of a glass vessel equipped with a diffuser connected to an $O_3$ generator. The oxygen flow was 2.0 L/min. Two different nominal $O_3$ concentrations of 30 and 200 ppm in $O_2$ were used for PR removal experiments.

For blanket photoresist and porous low-$k$ samples, Fourier-transform infrared spectroscopy (FTIR) and spectroscopic ellipsometry were used to characterize the film before and after modification by UV irradiation and subsequent wet clean. While for patterned samples, scanning electron microscopy (SEM) was used for evaluation of cleaning efficiency. The dielectric constant of blanket low-$k$ films was determined by Hg-probe.

## RESULTS AND DISCUSSION

Exposure of the PR film to the three-step plasma as described in the experimental part resulted in a substantial change of the PR film. Figure 1(a) shows the FTIR spectra of pristine and plasma-treated PR. The C=O absorption band attributed to the lactone (~1795 cm$^{-1}$) and ester group (~1730 cm$^{-1}$) drastically decreased in intensity after being subjected to the plasma process. With respect to the spectrum of pristine PR, the intensity ratio of these two bands is reversed, which gives a clear indication that the lactone groups are preferentially cleaved off. Likewise, the appearance of a small shoulder below 1700 cm$^{-1}$ was recorded. This shoulder is attributes to O-H group from $H_2O$ and C=C bonds. The UV ellipsometry spectra shown in Figure 1(b) confirm the FTIR data. In the case of poly methyl methacrylate (PMMA), it has been reported that C=C bonds were generated in the polymer chain as a result of the cleavage of the methyl ester side groups [12-13]. This cleavage can be followed by the scission of C-C bonds in the polymer chain. In more recent studies, plasma treatments for various times using a mixture of $O_2/CF_4/CH_2F_2$ performed on a PR of the same family as the one used in this study led to similar observations, where formation of C=C bonds along the polymer backbone from radical termination reaction was proposed [4,6].

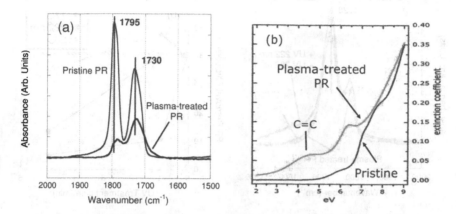

Figure 1: (a) FTIR and (b) Ultra-violet ellipsometry spectra of pristine and plasma-treated PR.

FTIR spectra recorded after UV irradiation for 1 min are shown in Figure 2(a). The absorption bands corresponding to lactone and ester groups that already drastically decreased due to the plasma treatment, further decreased in intensity after subsequent irradiation with 172-nm UV light. Almost no change was observed for the band attributed to C=C and OH groups. In contrast, irradiation by 222 nm as well as by 283 nm photons showed significant different results compared with 172 nm photons. The intensity of the shoulder assigned to C=C and OH groups significantly increased and became a clear broad band. The intensity of the lactone and ester bands was slightly higher, with similar lactone/ester ratio. While there were some qualitative differences in peak intensity that most likely resulted from a slight decrease in film thickness of the PR film after the UV treatment, the overall shape of the spectra and the lactone/ester peak ratio were remarkably similar for these two wavelengths. For both 222 nm and 283 nm UV-treated PR films, measurements using ellipsometry showed that the film thicknesses decreased only by about 2 nm after 1 min of treatment and remained constant for a 5-min treatment. Comparing to 172 nm photons, these results clearly indicate that treatments by 222 nm and 283 nm photons preserved better the lactone and ester groups of the molecule (less "degradation"), and in particular, they were more efficient for generation of C=C bonds with a little change in film thickness.

Figure 2(b) shows the change in removal efficiency as a function of UV treatment time evaluated by subsequent immersion of the PR film in NMP at 60°C for 2 min. Treatment by 172 nm UV source was found to be detrimental to PR removal, implying the formation of more crosslinked material, even for a short treatment time of 1 min. Similar behavior was also observed for the samples irradiated by 222 nm and 283 nm sources for a longer exposure time, i.e. 5 min. On the contrary, exposure to 283 nm photons for up to 1 min resulted in a slight improvement in removal efficiency. This result suggests that some chain scission occurred under these conditions resulting in smaller fragments and rendering it more soluble in the solvent.

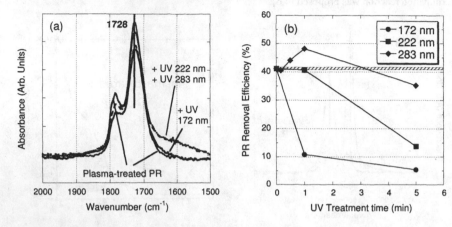

Figure 2: Effect of UV irradiation at different wavelengths applied on blanket plasma-treated PR, (a) FTIR data showing generation of C=C for 222 nm and 283 nm photons and (b) change in removal efficiency as a function of UV treatment time evaluated by subsequent immersion in NMP at 60°C for 2 min.

The results shown above give an evidence, at least for the treatment with 222 nm and 283 nm UV lamps, of a competition between chain scission and crosslinking, and formation of C=C bonds. The PR film treated by 222 nm UV source was chosen as a substrate for reaction with an oxidizer, $O_3$, as it led to a significant increase in C=C bonds. The SEM images in Figure 3 show a single damascene low-$k$ structure after different process steps. Exposure to $O_3/H_2O$ vapor at 90°C for 5 min followed by a rinse in organic solvent did not completely removed the PR and BARC layers on top of TiN hard mask (Figure 3(b)). An irradiation for 1 min with 222 nm photons prior to the exposure to $O_3/H_2O$ vapor under the same conditions resulted in a complete removal of both PR and BARC layers. The effect of UV treatment on BARC is believed to be due to a direct scission of the BARC molecule backbone. It is worth noticing that a complete removal of PR and BARC was only achieved with a rinse using an organic solvent. The fragments resulting from the reaction between $O_3$ and the molecular chain containing C=C bonds are most likely not small enough to be soluble in water.

The observed complete removal of PR layer when combining the UV treatment of 222 nm photons and exposure to an oxidizer such as $O_3$ is in turns supportive of the C=C formation due to UV treatment and consistent with the FTIR results shown in Figure 1.

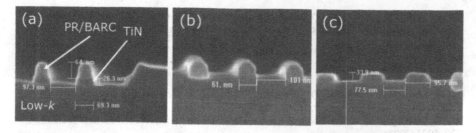

Figure 3: SEM images of single damascene low-$k$ structure, (a) after BARC and TiN etch, (b) with subsequent exposure to $O_3/H_2O$ vapor for 5 min and organic solvent rinse, (c) with 222 nm UV treatment prior to exposure to $O_3/H_2O$ vapor for 5 min and solvent rinse showing complete removal of PR and BARC layers

Experimental results obtained from characterization of the porous dielectric using FTIR and spectroscopic ellipsometry indicated that the dielectric film was little affected by UV irradiation by photons of the three wavelengths studied (Figure not shown). Also, $k$-value evaluation using Hg probe of a 283 nm-treated low-$k$, performed on a pristine low-$k$ substrate treated for 1 min, showed that the dielectric $k$-value remained very similar to that of the pristine one, i.e. ~2.55-2.60. This result is good agreement with previous studies by Prager et al., where the degradation of porous low-$k$ was found to be more pronounced for photons with wavelength below 200 nm [14]. However, $O_3/H_2O$ process did not seem to be fully compatible with this low-$k$ material. The $k$-value of the low-$k$ film showed an increase of about 0.3 after being processed with $O_3/H_2O$ for 5 min. In another series of experiment where, instead of water, $O_3$ was directly dissolved in PC resulting in a good compatibility of this process with the same porous low-$k$. This can be explained by the radical inhibitor role played by PC molecules. Reactive species such as oxygen (atomic or radical) or hydroxyl radicals are known to be detrimental to porous low-$k$ materials. In the case of $O_3$ dissolved in PC, radicals and other reactive species generated by the reaction with $O_3$ or by its decomposition immediately reacted with PC molecules at elevated temperature leading to formation of products such as acetic acid. A similar reaction was not possible in the case of $O_3/H_2O$.

## CONCLUSIONS

It has been shown that irradiation of post-etch PR by 283 nm photons for times up to 1 min led to bond scission that improved PR dissolution in organic solvent, while crosslinking seems to dominate under other conditions, especially for the other two wavelengths, 222 nm and 172 nm, for longer exposure times. There is an apparent competition between chain scission and crosslinking, together with formation of C=C bonds.

UV irradiation can be applied as an intermediate step to enhance PR wet strip using selective oxidizing chemistries. For PR removal purpose, the major benefit of UV treatments for PR wet strip consists in formation of C=C bonds. As a case study, the PR treated by 222 nm photons was chosen prior to reaction with an oxidizer, $O_3$ in $H_2O$ vapor, at 90°C. Experimental results showed significant removal improvement for UV pre-treated PR resulting in a complete removal of PR and BARC layers, which can be explained by C=C bond cleavage by the oxidizer.

The ability to choose UV wavelengths within the range of 222-283 nm for a treatment enabling complete PR and BARC removal gives considerably greater possibility to develop a valuable process.

## REFERENCES

1. K. Maex, M. R. Baklanov, D. Shamiryan, F. Iacopi, S. H. Brongersma, and Z. S. Yanovitskaya, J. Appl. Phys., 93, 8793 (2003).
2. D. Shamiryan, M. R. Baklanov, S. Vanhaelemeersch. and K. Maex, J. Vac. Sci. Technol. A, 20, 1923 (2002).
3. C. Waldfried, O. Escorcia, Q. Han, and P. B. Smith, Electrochem. Solid-St. Lett., 6, G137 (2003), and references therein.
4. M. Claes, Q. T. Le, E. Kesters, M. Lux, A. Urionabarrenetxea, G. Vereecke, and P. W. Mertens, ECS Trans., 11 (2) , 177 (2007).
5. Q. T. Le, , J. Keldermans, N. Chiodarelli, E. Kesters, M. Lux, M. Claes, and G. Vereecke, Jpn. J. Appl. Phys., 47, 6870 (2008).
6. E. Kesters, M. Claes, Q. T. Le, M. Lux, A. Franquet, G. Vereecke, P. W. Mertens, M.M. Frank, R. Carleer, P. Adriaensens, J. J. Biebuyck, S. Bebelman, Thin Solid Films, 516, 3454 (2008).
7. Q. T. Le, M. Claes, T. Conard, E. Kesters, M. Lux, and G. Vereecke, Microelectronic Eng., 86(2), 181 (2009).
8. S. De Gendt, J. Wauters, and M. M. Heyns, Solid State Technol., 41, 57 (1998).
9. F. De Smedt, S. De Gendt, M. M. Heyns, and C. Vinckier, J. Electrochem. Soc., 148, G. 487 (2001), and the references therein.
10. Q. T. Le, E. Kesters, L. Prager, M. Claes, M. Lux, and G. Vereecke, Solid State Phenom., 145-146, 323 (2009).
11. E. Kesters, M. Claes, Q. T. Le, K. Barthomeuf, M. Lux, G. Vereecke, and J. B. Durkee, Microelectronic Eng., 86(2), 160 (2009).
12. D. M. Ruck, J. Schulz, and N. Deusch, Nucl. Instr. Methods Phys. Res. B, **131**, 149 (1997).
13. P. Henzi, K. Bade, D. G. Rabus, and J. Mohr, J. Vac. Sci. Technol. B, **24** (4), 1755 (2006).
14. L. Prager, P. Marsik, J. W. Gerlach. M. R. Baklanov, S. Naumov, L. Pistol, D. Schneider, L. Wennrich, P. Verdonck, M. R. Buchmeiser, Microelectron. Eng., 85, 2094 (2008).

# Poster Session:
# Interconnects

Mater. Res. Soc. Symp. Proc. Vol. 1156 © 2009 Materials Research Society 1156-D03-01

# Effects of Silica Sources on Nanoporous Organosilicate Films Templated With Tetraalkylammonium Cations

Salvador Eslava,[1,2+] Jone Urrutia,[1] Abheesh N. Busawon,[3] Mikhail R. Baklanov,[1] Francesca Iacopi,[2] Karen Maex,[4] Christine E. A. Kirschhock,[2] and Johan A. Martens[2]

[1]IMEC, Kapeldreef 75, 3001 Leuven, Belgium. [2]Centrum voor Oppervlaktechemie en Katalyse, Katholieke Universiteit Leuven, Kasteelpark Arenberg 23, 3001 Leuven, Belgium. [3]Imperial College London, London, SW7 2AZ, UK. [4]Department of Electrical Engineering, Katholieke Universiteit Leuven, Kasteelpark Arenberg 10, 3001 Leuven, Belgium.[+]SE's current address: Chemistry Dept., Univ. of Cambridge, Lensfield Road CB2 1EW, UK (se296@cam.ac.uk).

## ABSTRACT

Nanoporous organosilicate films have been recently prepared using tetraalkylammonium cations in acid and basic media, outperforming other materials. Resulting films using basic medium were called zeolite-inspired low-$k$ dielectrics. Here we study the dependence of the properties of these films on the used silica sources: methyltrimethoxy silane (MTMS) and tetraethyl orthosilicate (TEOS). A set of experiments varying the MTMS:TEOS ratio were prepared in acid medium and characterized. A textural, physico-chemical, mechanical, and electrical characterization of this series of experiments is presented.

## INTRODUCTION

Applications of nanoporous organosilicates include different fields such as lasers, biotechnology, molecular separations, and low-$k$ dielectrics in integrated circuits.[1] For some of these applications, film deposition is required. The pore network and the matrix composition are crucially important for the final film properties.[2] To obtain a porous network, several studies have been carried out involving the use of sacrificial molecules called porogens.[1-4] Recently, we have shown that tetraalkylammonium bromides (TAABr) of different alkyl chain length outperform other porogen molecules in terms of porosity results.[5] This was demonstrated by using TAABr molecules as porogens in the spin coating of organosilicate sols prepared from an equimolar quantity of tetraethyl orthosilicate (TEOS) and methyltrimethoxysilane (MTMS) in acid conditions. The alkyl chain length of the TAABr and the porogen concentration defined the final pore network. Other properties such as hydrophobicity, elastic modulus, and dielectric constant were shown to be dependent on the pore network nanocasted by the TAABr molecules. Alternatively, organosilicate films could also be prepared in basic conditions using aqueous tetraalkylammonium hydroxides (TAAOH), giving rise to zeolite-inspired low-$k$ (ZLK) dielectrics.[6] The comparison of obtained films with other materials such as spin-on zeolite films evidenced the advantages of these approaches.[5,6]

In this work, we contribute to the study on the organosilicate films nanocasted with TAABr porogens. We study the dependence of the final film properties on the proportion of MTMS relative to TEOS. A series of experiments varying the MTMS:TEOS ratio and keeping rest of variables were prepared and characterized. In the preparation, tetrapropyl-n-ammonium bromide (TPABr) porogen with a ratio TPABr/Si equal to 0.1 was chosen as a fixed variable.

## EXPERIMENTAL

An organosilicate sol was prepared following same recipe described in [5], in which tetraethyl orthosilicate (TEOS, 98% purity, Acros Organics) and methyltrimethoxy silane (MTMS, 98%, Fluka) were partially hydrolyzed and condensed under stirring at 60 °C for 90 min. in presence of ethanol (EtOH), doubly deionized (DDI) water, and HCl. In this work different ratios of MTMS:TEOS were adopted. The initial molar ratio of that step was $x$MTMS: $(1-x)$TEOS: 0.88water: 0.00005HCl: 3EtOH, where $x$ was varied between 0, 0.25, 0.5, and 0.75. In a second step, it was added a mixture of ethanol, water, HCl, and TPABr and stirred for 3 days. The final molar ratio was $x$SiO$_{1.5}$CH$_3$: $(1-x)$SiO$_2$: $(3+0.5x)$water: 0.004HCl: 20EtOH: $3x$Methanol: 0.1TPABr. The final pH was in the range 2-4. The sols were then filtered through 200 nm PTFE filters before deposition. Films were deposited on 3x3 cm$^2$ pieces of Si-p and Si-n$^{++}$ wafers by spin coating at 2,000 rpm. The films were first baked at 150 °C for 1 min. and then calcined in air at 400 °C for 30 min.

The chemical bonding structure of the films was characterized by FT IR spectroscopy (Biorad FTS-40) in nitrogen atmosphere in transmission mode. Spectra were normalized by the Si-O-Si antisymmetric stretching integrated in the region 970-1250 cm$^{-1}$ upon subtracting the Si-CH$_3$ related absorption band. Thickness and refractive index was measured by spectroscopic ellipsometry in the region 350-850 nm (Sentech SE801). Water and toluene adsorption isotherms and pore radius distributions were obtained by ellipsometric porosimetry (EP).[7] The dielectric constant of the calcined films was measured at 100 kHz by impedance analysis (HP4284A-LCR meter). The elastic modulus and hardness were measured by nanoindentation in continuous stiffness measurement with a Berkovich tip for films thicker than 850 nm.

## RESULTS AND DISCUSSION

The FT IR spectra for calcined films prepared from different MTMS:TEOS ratio are shown in Figure 1, showing common IR absorption bands of silicates.[5] Films prepared from higher MTMS fraction had higher IR absorption related to SiCH$_3$ groups (~2980, ~1270, and 850-750 cm$^{-1}$). On the other hand, higher concentration of MTMS led to a decrease of the O-H (3800-3300 cm$^{-1}$) and Si-OH stretching (950 cm$^{-1}$). Finally, films prepared from higher MTMS content showed a Si-O-Si antisymmetric stretching (AS) with a more pronounced shoulder at higher frequencies (centered at ~1140 cm$^{-1}$) as well a broadening of the AS at 1030-1080 cm$^{-1}$ towards lower frequencies. The increase of the IR band at ~1140 cm$^{-1}$ is commonly attributed to the formation of cagelike domains typical for organosilicate films.[5] The broadening is attributed to the presence of Si-O-Si angles smaller than 144° caused by the linkage of CH$_3$ groups to one of the Si atoms.[8]

The porosity in the films was characterized by EP. At first, the porosity was calculated using the refractive index in high vacuum atmosphere and applying Lorentz-Lorentz equation (Table 1).[7] The refractive index of the skeleton was approximated to that of fused silica (1.46). The incorporation of organic moieties in the skeleton should increase slightly the refractive index of the skeleton, so considering the skeleton to be pure silica gave rise to a minimum value of the porosity. Extra experiments (not shown) on similar materials taught that the error margin would not be higher than 4 vol.-%. Porosity varied between 27 and 35 vol.-%, tending to increase with the use of MTMS in the preparation. Toluene adsorption uptakes measured at 633 nm wavelength are shown in Figure 2. The maximum toluene uptake was 31-34 vol.-%, close to the values derived from the refractive index (Table 1). The resulting average pore radii and pore radius distributions are shown in Table 1 and Figure 2b, resp. The pores were larger in films prepared from higher ratio of MTMS. Larger pore radius and porosity in organosilicate films upon use of MTMS are explained considering the chemical bonding structure analyzed by FT IR. A higher MTMS fraction in silica sources decreases the chances to obtain a volume spanning cross-linked skeleton. Each hydrolysis of TEOS resulted in 4 Si-OH moieties that contributed to the silicate cross-linking by silanol condensation. However, each MTMS contributed only 3 Si-OH moieties. The remaining Si-CH$_3$ group could not contribute to the cross-linking since Si-CH$_3$ groups are not prone to hydrolysis.[9] Moreover, they occupy larger volumes than oxygen, creating considerable spatial hindrance. This promotes an increased pore size and inherent porosity in the films with increasing content of MTMS.

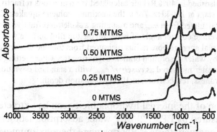

**Figure 1.** FT IR absorbance in the region 3900-400 cm$^{-1}$ of calcined organosilicate films prepared with different MTMS fraction in silica sources.

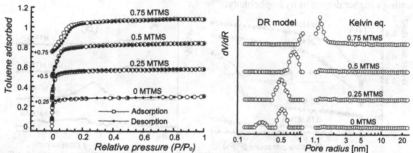

**Figure 2.** (a) Toluene adsorption isotherms from EP on calcined organosilicate films prepared with different MTMS fraction in silica sources. Y axis shows the volume fraction of toluene ($vol_{toluene}/vol_{film}$). The origin of each isotherm has been shifted by the indicated values to avoid overlap. (b) Pore radius distribution measured from toluene desorption branches, analyzed by Dubinin-Raduskevitch model and Kelvin equation.

Since the presence of water in the films jeopardizes the application of these films as low-$k$ dielectrics, hydrophobicity is an important parameter. The results of water contact angle measurements on films are presented in Table 1. With increasing MTMS fraction the contact angle increased from 33($\pm$2) ° to 91($\pm$4) °. These values describe hydrophobicity of the top surface of the films. To investigate the internal hydrophobicity, spectroscopic ellipsometry in function of humidity allows quantification of the adsorbed water. The refractive index of the films was measured at 40 % relative humidity (RH) in cleanroom air and upon degassing at high vacuum (Table 1). Thus, the calculated percent of water present in the film at 40% RH prepared without MTMS was 25 vol.-%. This value decreased to 0.6 vol.-% upon use of MTMS (Table 1).

**Table 1.** Properties of the calcined organosilicate films prepared with different types of TAA molecules and different TAA/Si molar ratios*

| MTMS/Si | $n_{air}$ | $n_{vac}$ | $W_{40\%RH}$ [vol.-%] | $k$ | $Tol$ [vol.-%] | $P$ [vol.-%] | $CA_{surf}$ [deg] | $r$ [nm] | $r_{wat}$ [nm] | $CA_{int}$ [deg] | $E$ [GPa] | $H$ [GPa] |
|---|---|---|---|---|---|---|---|---|---|---|---|---|
| 0 | 1.419 | 1.324 | 25 | >10 | 31 | 27 | 33 | 0.25,0.5 | 0.7 | 45 | 16.5 | 1.59 |
| 0.25 | 1.365 | 1.297 | 19 | 9.5 | 33 | 32 | 68 | 0.5 | 0.9 | 56 | 9.7 | 1.18 |
| 0.5 | 1.312 | 1.305 | 1.9 | 2.8 | 34 | 31 | 92 | 0.8 | 4.5 | 79 | 5.5 | 0.80 |
| 0.75 | 1.287 | 1.285 | 0.6 | 2.5 | 33 | 35 | 91 | 1.25 | ~13 | 85 | 3.7 | 0.45 |

*$n_{air}$ is the refractive index at 633 nm in 40 % RH air atmosphere; $n_{vac}$ is the refractive index at 633 nm in high vacuum; $W_{40\%RH}$ is the water adsorbed in 40% RH air calculated from previous refractive indices by Lorentz-Lorentz equation; $k$ is the dielectric constant at 100 kHz; $Tol$ is the maximum toluene uptake measured by EP with 633 nm wavelength; $P$ is the porosity calculated from $n_{vac}$ and assuming a skeleton refractive index of 1.46 (approx. to that of fused silica); $CA_{surf}$ is the average contact angle of water on the top film surface with a standard deviation smaller than 4 deg in all cases; $r$ is the real average pore radius measured from toluene desorption branches; $r_{wat}$ is the average pore radius measured by water desorption branches ignoring the water wetting angle ($r_{wat}=rcos\Theta$); and $CA_{int}$ is the internal contact angle of water measured as $cos\Theta=r/r_{wat}$, with a standard deviation of approx. 10 °. $E$ and $H$ are the elastic modulus and hardness, respectively, at 10 % indentation depth.

Further information of the internal hydrophobicity can be extracted from water adsorption isotherms measured by EP.[8,10,11] EP water adsorption isotherms are presented in Figure 3. Best features to compare these isotherms were the maximum water uptake and the slope of the adsorption branch at low pressures. Both decreased with the use of MTMS indicating a higher degree of hydrophobicity.[10]

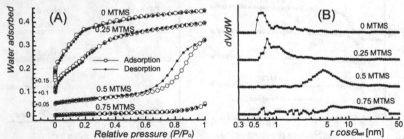

**Figure 3.** (a) Water adsorption isotherms determined by EP on calcined organosilicate films prepared with different MTMS fraction in silica sources. The origin of each isotherm was shifted by the indicated values to avoid overlap. (b) Pore radius distribution measured from previous water desorption branches ignoring wetting angle of water using Kelvin equation.

The comparison between the pore radius measurements obtained from toluene and water adsorption isotherms permits to calculate the internal contact angle of water in porous films.[10] This measurement relies on the validity of the Kelvin equation for the adsorption of different adsorbates. The Kelvin equation is written: $r=-2\gamma V_L \cdot cos\Theta \cdot R^{-1} T^{-1} (-lnP/P_o)$, where $r$ is the Kelvin radius, $\gamma$ is the surface tension, $V_L$ is the molar volume of the liquid adsorptive, and $\Theta$ is the contact angle of the adsorptive. Toluene has excellent wetting properties, which has made it an optimal adsorbate for the EP pore radius analysis. The contact angle of toluene is always assumed to be zero, so the term $cos\Theta$ can be neglected. Thus, the radius measured with toluene adsorbate ($r_{tol}$) is equal to the real pore radius ($r$). On the other side, the wetting properties of water depend on the hydrophilicity of the internal pore surface of the films. If the film is very hydrophilic, the contact angle of water would be zero and thus would lead to a pore radius distribution similar to that obtained by toluene ($r_{wat}=r_{tol}=r$). However, if the contact angle of water is larger than zero, the pore radius distribution obtained with water adsorbate is not reflecting the real pore radius weighed by the cosine of the contact angle ($r_{wat} = r \cdot cos\Theta_{wat}$). This allows the calculation of the cosine of the internal contact angle of water as ratio of $r_{tol}/r_{wat}$.

The pore radius distribution obtained with water adsorbate is shown in Figure 3b. The average pore radius obtained with water adsorbate and the internal contact angle are summarized in Table 1. The surface contact angle was measured to be between 33($\pm$2) ° and 91($\pm$4) ° depending on the MTMS ratio. Internal or pore surface contact angle were between 45($\pm$10) ° and 85($\pm$10) °. It increased with the ratio of MTMS in agreement with the surface contact angle and the shape of the isotherms. Considering standard deviations, the internal and top film contact angles indicated similar hydrophobicity.

The differences in the hydrophilicity stemmed from the different type of terminating groups in the organosilicate matrix. TEOS and MTMS sources could give rise to hydrophilic terminating groups (Si-OH) upon vitrification of the matrix but MTMS could in addition add hydrophobic terminating groups (Si-CH$_3$). Thus, in absence of MTMS in the synthesis, the hydrolysis and condensation of TEOS led to pore-surfaces covered by silanols and water as observed in FT IR analysis (Fig. 1). Full silanol condensation during the heating step was not achieved at 400 °C. Alternatively, ultraviolet light can be used to enhance the silanol condens-ation at lower temperatures (<450 °C) but it was not adopted in this series of experiments.[11] When a fraction of TEOS source was substituted by MTMS in the synthesis, Si-CH$_3$ terminating groups were present during the hydrolysis and condensation process. The stability of these groups is known in literature, so they could persist during whole process.[9] The presence of terminating Si-CH$_3$ in the final matrix offered repulsive sites for water adsorption and in turn, decreased the quantity of hydrophilic Si-OH moieties, as observed by FT IR (Fig. 1). In absence of Si-OH moieties, the hydrophobicity was ensured because the pore walls could not be covered by water films enabling capillary condensation. Coupled to this, the larger pore radii obtained in films prepared from higher MTMS ratio (Table 1) could also contribute to higher hydrophobicity in films, although with minor importance. This is due to the increased difficulty to physisorb water by hydrogen bonding in flatter pore surfaces found in larger pores.[5]

The mechanical properties are of interest for the application of these films in integrated circuits, and were measured by nanoindentation (Table 1). With the use of MTMS, the elastic modulus and hardness decreased from ~16 to ~3 GPa and from ~1.6 to ~0.4 GPa, respectively. The strong dependence could not be attributed exclusively to the slight changes of the porosity in the films but also to the different matrix cross-linking. As previously described, Si-CH$_3$ moieties resulting from use of MTMS can be considered as networkbreakers reducing the cross-linking of the silica framework. The networkbreaking properties of Si-CH$_3$ groups were confirmed in FT IR spectra on the region of C-H$_x$ stretching of alkyl bridges, Si-CH$_2$-CH$_2$-Si. There was only very weak absorption of C-H$_2$ species at ~2915 cm$^{-1}$ (Figure 1), indicating minimal presence, if any, of alkyl bridges in the network. In a nutshell, the Si-O-Si matrix obtained with lower MTMS fraction had longer chains, higher cross-linking, and this favored the resistance to withstand pressure.

The dielectric constant of films prepared with different MTMS fraction was measured at 100 kHz and listed in Table 1. Dielectric constants were higher than 10 in absence of MTMS in the reactant mixture but decreased down to ca. 2.5 upon use of MTMS. Hydrophilic films (0 and

0.25 MTMS films) showed a dielectric constant higher than that of nonporous silica ($k_{SiO_2}$~4). The porosity decreased the $k$ value due to the low dielectric constant of gases (~1) in the pore network but this was not enough to compensate the presence of physisorbed water with a dielectric constant ca. 78. Note that at 40 % RH the physisorbed water was as high as 25 and 23 vol.-% (*vide supra*). Chemisorbed water, *i.e.*, Si-OH moieties with high polarizability, also contributed significantly to increase the dielectric constant. Films prepared with 0.5 and 0.75 MTMS fraction were hydrophobic, so the pore network contained minimal quantities of condensed water resulting in lower dielectric constants. In addition, the use of MTMS decreased the concentration of Si-O bonds forming part of SiOH and SiOSi groups and increased the concentration of Si-CH₃. The lower polarizability of Si-CH₃ in comparison to Si-O bonds decreased the high frequency dispersion occurring in silica-based materials which in turn decreased the dielectric constant at 100 kHz.

## CONCLUSIONS

TEOS and MTMS were used in different ratios for the preparation of nanoporous organosilicate films prepared with TPABr porogen. The synthesis and characterization of this set of experiments demonstrated the strong effects of the MTMS:TEOS ratio on the final properties of the films. Increasing MTMS fraction led to increase the quantity of organic moieties (Si-CH₃) and in turn the quantity of silanols in the films decreased. The organic moieties caused spatial hindrance during film formation increasing porosity and pore size. The decrease of silanol moieties improved hydrophobicity in the films. On the other hand, the presence of organic moieties decreased the cross-linking of the silicate matrix deteriorating the film mechanical properties elastic modulus and hardness. The dielectric constant decreased mainly as a result of the reduction of polar OH groups.

## ACKNOWLEDGMENTS

CEAK and JAM acknowledge the Flemish government for a concerted research action (GOA).

## REFERENCES

1. G. Dubois, R.D. Miller, W. Volksen, in *Dielectric Films for Advanced Microelectronics*, (Eds: M.R. Baklanov, M. Green, K. Maex), John Wiley & Sons Inc, England **2007**, Ch. 2.
2. F.K. de Theije, A.R. Balkenende, M.A. Verheijen, M.R. Baklanov, K.P. Mogilnikov, Y. Furukawa, *J. Phys. Chem. B* **2003**, *107*, 4280.
3. M.H. Ree, J.W. Yoon, K.Y. Heo, *J. Mater. Chem.* **2006**, *16*, 685.
4. L. Nicole, C. Boissiere, D. Grosso, A. Quach, C. Sanchez, *J. Mater. Chem.* **2005**, *15*, 3598.
5. S. Eslava, M.R. Baklanov, J. Urrutia, C.E.A. Kirschhock, K. Maex, J.A. Martens, *Adv. Funct. Mater.* **2008**, *20*, 3110.
6. S. Eslava, J. Urrutia, A. N. Busawon, M. R. Baklanov, F. Iacopi, S. Aldea, K. Maex, J. A. Martens, C. E. A. Kirschhock, *J. Am. Chem. Soc.* **2008**, *130*, 17528.
7. S. Eslava, M.R. Baklanov, C.E.A. Kirschhock, F. Iacopi, S. Aldea, K. Maex, J.A. Martens, *Langmuir* **2007**, *23*, 12811.
8. S. Eslava, S. Delahaye, M.R. Baklanov, F. Iacopi, C.E.A. Kirschhock, K. Maex, J.A. Martens, *Langmuir* **2008**, *24*, 4894.
9. R.H. Baney, M. Itoh, A. Sakakibara, T. Suzuki, *Chem. Rev.* **1995**, 95, 1409.
10. M.R. Baklanov, K.P. Mogilnikov, Q.T. Le, *Microelectron. Eng.* **2006**, *83*, 2287.
11. S. Eslava, F. Iacopi, M.R. Baklanov, C.E.A. Kirschhock, K. Maex, J.A. Martens, *J. Am. Chem. Soc.* **2007**, *129*, 9288.

Mater. Res. Soc. Symp. Proc. Vol. 1156 © 2009 Materials Research Society

# Electrical and Structural Properties of Ultrathin Polycrystalline and Epitaxial TiN Films Grown by Reactive dc Magnetron Sputtering

F. Magnus[1], A. S. Ingason[1], S. Olafsson[1], and J. T. Gudmundsson[1,2]

[1]Science Institute, University of Iceland, Dunhaga 3, IS-107 Reykjavik, Iceland

[2]Department of Electrical and Computer Engineering, University of Iceland, Hjardarhaga 2-6, IS-107 Reykjavik, Iceland

## ABSTRACT

Ultrathin TiN films were grown by reactive dc magnetron sputtering on amorphous $SiO_2$ substrates and single-crystalline MgO substrates at 600°C. The resistance of the films was monitored *in-situ* during growth to determine the coalescence and continuity thicknesses. TiN films grown on $SiO_2$ are polycrystalline and have coalescence and continuity thicknesses of 8 Å and 19 Å, respectively. TiN films grow epitaxially on the MgO substrates and the coalescence thickness is 2 Å and the thickness where the film becomes continuous cannot be resolved from the coalescence thickness. X-ray reflection measurements indicate a significantly higher density and lower roughness of the epitaxial TiN films.

## INTRODUCTION

TiN thin films are widely used as adhesion layers and diffusion barriers in Cu interconnects in modern microelectronics [1]. This is in part due to the excellent thermal stability and low bulk electrical resistivity of TiN. Recently, TiN has been proposed as a direct-metal-gate material for metal-oxide-semiconductor devices [2]. For the above applications it is desirable to reduce the thickness of the TiN film as much as possible but the film must remain continuous. In addition, it is important to maintain the low resistivity even in the ultrathin limit. Furthermore, highly textured or epitaxial thin MgO films have received much interest lately due to their applications in magnetic tunnel junctions [3]. Epitaxial MgO films could also be a possible candidate for an intermediate-to-high-κ gate dielectric [4]. It is therefore important to find a suitable gate-metal for the MgO gate dielectric.

Here we compare the electrical and structural properties of ultrathin TiN films grown on $SiO_2$ substrates and single-crystalline MgO[200] substrates by reactive dc magnetron sputtering. The electrical resistance of the films was monitored during growth *in-situ* to determine the coalescence thickness and the thickness at which the films become continuous. In addition, we examine crystallinity, density and roughness of the films *ex-situ* with X-ray diffraction (XRD) and reflection (XRR) measurements.

## EXPERIMENT

The TiN thin films were grown in a custom built magnetron sputtering chamber [5] with a base pressure of $4 \times 10^{-6}$ Pa for 30 minutes. The sputtering gas was argon of 99.999% purity mixed with nitrogen gas of 99.999% purity. The argon flow rate was 40 sccm and the nitrogen flow rate 2 sccm to give a pressure of 0.4 Pa. The Ti target was 50 mm in diameter and of 99.99% purity. The applied power was set to 100 W. The substrates used were thermally oxidized Si[001] with an oxide thickness of 500 nm and single-crystal MgO[200]. Au contact pads were defined on the substrates prior to TiN deposition. The substrate temperature was controlled during growth with a plate heater, separated from the substrate holder by a 2 mm gap [5]. The differential resistance of the TiN film was measured in a four-terminal configuration during growth using a simplified version of the dual lock-in amplifier setup described by Barnat et al. [6] as discussed in detail elsewhere [5]. The nominal coalescence thickness was determined by finding the maximum of $Rd^2$ vs. the film nominal thickness $d$, where $R$ is the measured film resistance (or equivalently $Rt^2$ vs. growth time $t$). The nominal film thickness which completely covers the substrate was determined by the minimum of $Rd^2$ vs. $d$ [7,8].

X–ray measurements were carried out using a Philips X'pert diffractometer (Cu $K_\alpha$, wavelength 0.15406 nm) mounted with a hybrid monochromator/mirror on the incident side and a 0.27° collimator on the diffracted side. The film thickness was determined by low–angle X–ray reflectivity (XRR) measurements with an angular resolution of 0.005°. High–angle θ–2θ scans were used for structural characterization of the films grown on a MgO substrate whereas grazing incidence (GI) XRD was used for films grown on $SiO_2$ to eliminate the substrate contributions. The GI scans were carried out with the incident beam at θ = 1°. In addition, a 360° scan of the azimuthal out of plane angle was performed at the TiN [111] peak to investigate film texture. The growth rate was determined by growing a series of films over a range of thicknesses. The growth rate was found to be 0.0227 nm/s for films grown on $SiO_2$ and 0.0215 nm/s for films grown on MgO. However, it should be kept in mind that the thickness of a discontinuous film is a somewhat ambiguous term and therefore we will often refer to the nominal thickness which is based on the constant growth rate.

## RESULTS AND DISCUSSION

The electrical resistance $R$ of TiN films grown on $SiO_2$ and MgO substrates at 600 °C, measured *in-situ* during growth, is shown in figure 1. Initially, the films are discontinuous as evidenced by the high resistance plateau. As the first conducting link forms across the surface the resistance drops sharply, which is known as the coalescence threshold [8]. This has been shown to correspond to the local maximum in the $Rd^2$ curve shown in the inset to figure 1. Immediately after coalescence, the films are electrically continuous but still have voids through to the substrate. The nominal thickness at which the voids have been filled in and the film becomes structurally continuous has been shown to correspond to the local minimum in the $Rd^2$ curve.

TiN has a cubic NaCl-type crystal structure with a lattice constant of 4.24 Å. TiN is expected to grow in three-dimensional islands on $SiO_2$ substrates [9]. The inset to figure 1 has a distinct maximum and minimum and we determine the coalescence thickness to be $8 \pm 1$ Å and the continuity thickness $19 \pm 2$ Å or the equivalent of approximately four atomic layers.

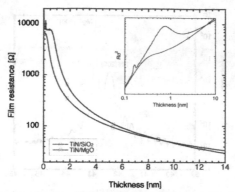

**Figure 1.** The film resistance $R$ as a function of TiN nominal thickness $d$ measured *in-situ* during growth, on a $SiO_2$ substrate and single-crystal MgO. The growth temperature is 600 °C. The thickness is based on the constant growth rates for TiN on $SiO_2$ and MgO. The inset shows $Rd^2$ versus thickness which is used to determine the coalescence and continuity thicknesses.

In contrast, MgO has the same crystal structure as TiN and a lattice constant of 4.21 Å. For perfect epitaxial growth, coalescence will occur before a full atomic layer has been deposited and the continuity thickness is exactly one atomic layer. Indeed we see from the onset of conductance in figure 1 that coalescence of the TiN film on MgO occurs at a nominal thickness of $2 \pm 1$ Å, or less than a single atomic layer. However, we cannot resolve the coalescence and continuity thicknesses from the $Rd^2$ curve as they occur at almost the same thickness.

Earlier, we have shown that thin polycrystalline TiN films grown on $SiO_2$ substrates at 600 °C are stoichiometric, have a high density and a low resistivity of $54 \pm 4$ μΩ cm [10]. We find that the resistivity of TiN films grown on MgO is significantly lower, or $24 \pm 4$ μΩ cm. The resistivity of a polycrystalline film is influenced by grain boundary scattering which is not present in epitaxially grown films. In addition, an epitaxial film can be expected to have lower surface and interface roughness than a polycrystalline film and therefore a lower resistivity contribution from diffuse surface/interface scattering. Indeed, the low surface roughness is confirmed by the XRR measurements presented below. The resistivity and coalescence and continuity thicknesses are summarized in table 1.

High-angle XRD scans were made of the TiN films grown on MgO to determine the crystal structure. The scan of the TiN [200] peak is shown in figure 2. The MgO substrate peak is at 42.895° which by Bragg's law gives a lattice constant of 4.21 Å and the central TiN peak is at 42.349° which corresponds to a lattice constant of 4.27 Å. Well resolved Laue-oscillations are also visible on both sides of the TiN Bragg peak indicating good epitaxial growth of the TiN. The thickness of the TiN film can be determined from the period of the oscillations by

$$d = \frac{\lambda}{2(\sin\theta_{i+1} - \sin\theta_i)} \tag{1}$$

where $\lambda$ is the wavelength, and $\theta_i$ and $\theta_{i+1}$ are two adjacent maxima or minima [11]. By this method we determine the TiN thickness as 37.5 nm.

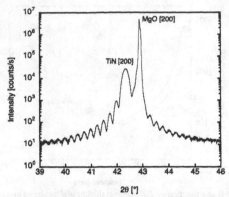

**Figure 2.** A high-angle XRD pattern of the TiN [200] peak for TiN grown on a MgO substrate at 600 °C. Laue oscillations can clearly be seen, indicating good epitaxial growth of the TiN.

GI-XRD measurements of the TiN/SiO₂ samples grown at various temperatures are shown in figure 3. The TiN films are clearly polycrystalline and the [111], [200] and [220] orientations are all present in the samples grown at 600 °C. A change in the dominant crystal orientation from [200] to [111] has been shown to occur with increasing thickness [12] but our films are below this cross-over thickness. The [200] peak is the most prominent at 600 °C but recedes with decreasing growth temperature and is not present in room temperature grown films. This growth temperature dependence of preferred orientation in polycrystalline TiN thin films is consistent with previous studies and can be attributed to the competition between the temperature dependent fast lateral growth rate of the [200] oriented grains and the large geometric growth rate of the [111] oriented grains [12].

The average grain size is calculated from the full width at half-maximum of the GI-XRD peaks using the Scherrer formula [13] and is found to be approximately 30 nm in the 600 °C grown films and drop with decreasing growth temperature. The reduced grain size at lower growth temperatures results in increased grain boundary scattering and hence an increase in resistivity. Furthermore, the surface area of the TiN grains is increased, making the film more susceptible to oxidation [10].

**Table 1.** Properties of TiN films grown on SiO₂ and MgO at 600 °C for 30 min. The density, thickness and roughness is determined by the XRR measurements shown in figure 4. For the TiN film grown on SiO₂ the density refers to the TiN layer, the thickness is the total deposited thickness (TiN and TiO$_x$N$_y$ layers) and the roughness is the surface roughness. The coalescence and continuity thicknesses are determined from the in-situ resistance measurements shown in figure 1.

| Substrate | Density (g/cm³) | Thickness (nm) | Roughness (nm) | Resistivity (μΩ cm) | Coalescence thickness (Å) | Continuity thickness (Å) |
|-----------|-----------------|----------------|----------------|---------------------|---------------------------|--------------------------|
| SiO₂ | 4.9 | 41.7 | 2.7 | 54 ± 4 | 8 ± 1 | 19 ± 2 |
| MgO | 5.1 | 38.7 (37.5*) | 1.0 | 24 ± 4 | 2 ± 1 | – |

* From Laue oscillations

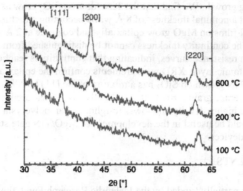

**Figure 3.** GIXRD scans of the TiN films grown on $SiO_2$ for 30 min. at different temperatures, shown on the right. The curves are shifted for clarity.

XRR analysis was carried out to determine the film thickness, density and roughness by curve fitting based on the Parratt formalism for reflectivity [14]. The results can be seen in figure 4. There is a clear difference between the films grown on $SiO_2$ and MgO. The TiN/$SiO_2$ is best fitted by a two-layer model including a 5.6 nm thick surface oxynitride layer whereas the TiN/MgO sample can be fitted with a single TiN layer. The fitting parameters are summarized in table 1. As expected, the epitaxial TiN on MgO has a higher density than the polycrystalline TiN on $SiO_2$, close to the theoretical value of 5.4 g/cm$^3$ [15]. Correspondingly, it also has a 7% lower thickness and significantly lower roughness.

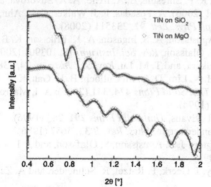

**Figure 4.** XRR measurements of the TiN films grown on $SiO_2$ (circles) and MgO (diamonds) at 600 °C. The solid lines are fits based on the Parratt formalism for reflectivity [14].

## CONCLUSIONS

Ultrathin TiN films have been grown by reactive dc magnetron sputtering on both amorphous $SiO_2$ substrates and single-crystalline MgO substrates. The resistance of the films has been monitored during growth. We find that the TiN films on $SiO_2$ grow as three-dimensional islands and coalesce at a nominal thickness of 8 Å while becoming structurally continuous at 19 Å. In contrast, the TiN films on MgO grow epitaxially and coalesce at 2 Å or below the thickness of one atomic layer. The continuity thickness cannot be distinguished from the coalescence thickness in the in-situ resistance curves, indicating that continuity is reached very close to the thickness of a single atomic layer. XRD measurements confirm the epitaxial growth of TiN [200] on MgO whereas the TiN grown on $SiO_2$ has a mixture of the [111], [200] and [220] crystal orientations with an average grain size of 30 nm. Finally, XRR measurements show that epitaxial TiN grown MgO has a higher density and lower roughness than polycrystalline TiN grown on $SiO_2$. These results could be useful in the development of MgO/TiN gate stacks for future metal-oxide-semiconductor devices.

## ACKNOWLEDGMENTS

This work was partially funded by the Icelandic Research Fund, the University of Iceland Research fund and the Steinmaur foundation.

## REFERENCES

1.  J. D. Plummer, M. D. Deal, and P. B. Griffin, *Silicon VLSI Technology: Fundamentals, Practice and Modeling.* (Prentice Hall, New Jersey, 2000).
2.  R. Chau, S. Datta, M. Doczy, B. Doyle, J. Kavalieros, and M. Metz, *IEEE Electron Device Lett.* **25**, 408 (2004); J. Westlinder, T. Schram, L. Pantisano, E. Cartier, A. Kerber, G. S. Lujan, J. Olsson, and G. Groeseneken, *IEEE Electron Device Lett.* **24**, 550 (2003).
3.  S. Yuasa, *J. Phys. Soc. Jpn.* **77**, 031001 (2008).
4.  L. Yan, C. M. Lopez, R. P. Shrestha, E. A. Irene, A. A. Suvorova, and M. Saunders, *Appl. Phys. Lett.* **88**, 142901 (2006); A. Posadas, F. J. Walker, C. H. Ahn, T. L. Goodrich, Z. Cai, and K. S. Ziemer, *Appl. Phys. Lett.* **92**, 233511 (2008).
5.  U. B. Arnalds, J. S. Agustsson, A. S. Ingason, A. K. Eriksson, K. B. Gylfason, J. T. Gudmundsson, and S. Olafsson, *Rev. Sci. Instrum.* **78**, 103901 (2007).
6.  E. V. Barnat, D. Nagakura, and T. M. Lu, *Rev. Sci. Instrum.* **74**, 3385 (2003).
7.  F. A. Burgmann, S. H. N. Lim, D. G. McCulloch, B. K. Gan, K. E. Davies, D. R. McKenzie, and M. M. M. Bilek, *Thin Solid Films* **474**, 341 (2005); A. I. Maaroof and B. L. Evans, *J. Appl. Phys.* **76**, 1047 (1994).
8.  I. M. Rycroft and B. L. Evans, *Thin Solid Films* **291**, 283 (1996).
9.  E. Bauer and J. H. Vandermerwe, *Phys. Rev. B* **33**, 3657 (1986).
10. A. S. Ingason, F. Magnus, J. S. Agustsson, S. Olafsson, and J. T. Gudmundsson, *Thin Solid Films* (submitted 2009).
11. G. Linker, R. Smithey, J. Geerk, F. Ratzel, R. Schneider, and A. Zaitsev, *Thin Solid Films* **471**, 320 (2005).
12. S. Mahieu and D. Depla, *J. Phys. D* **42**, 053002 (2009).
13. M Birkholz, *Thin film analysis by X-ray scattering.* (Wiley-VCH, Weinheim, 2006).
14. L. G. Parratt, *Phys. Rev.* **95**, 359 (1954).
15. H. Holleck, *J. Vac. Sci. Technol. A* **4**, 2661 (1986).

Mater. Res. Soc. Symp. Proc. Vol. 1156 © 2009 Materials Research Society    1156-D03-07

# A Study of Diffusion Barrier Characteristics of Electroless Co(W,P) Layers to Lead-Free SnAgCu Solder

Hung-Chun Pan and Tsung-Eong Hsieh[1]
Department of Materials Science and Engineering, National Chiao Tung University,
1001 Ta-Hseuh Road, Hsinchu, Taiwan 30010, R.O.C.
[1]Correspondence author's e-mail: tehsieh@mail.nctu.edu.tw

## ABSTRACT

Diffusion barrier characteristics of amorphous and polycrystalline electroless Co(W,P) layers ($\alpha$-Co(W,P) and poly-Co(W,P)) to lead-free SnAgCu (SAC) solder were investigated *via* the liquid- and solid-state aging tests. In the sample containing $\alpha$-Co(W,P) subjected to liquid-state aging at 250°C for 1 hr, the spallation of $(Co,Cu)Sn_3$ intermetallic compound (IMC) into the solder and formation of a polycrystalline P-rich layer in between SAC and Co(W,P) were found. Further, the $\alpha$-Co(W,P) transforms into polycrystalline structure embedded with tiny $Co_2P$ precipitates. As to the sample containing $\alpha$-Co(W,P) subjected to solid-state aging at 150°C up to 1000 hrs, a thick $(Cu,Co)_6Sn_5$ IMC resided in between SAC and Co(W,P) and the P-rich layer beneath IMCs was similarly seen. In the samples containing poly-Co(W,P) subjected to liquid-state aging, interactions of SAC and Co(W,P) implied a mixture of $(Co,Cu)Sn_3$ and $(Co,Ag)Sn_3$ IMCs and an amorphous W-rich layer formed in between IMCs and poly-Co(W,P). Similar interfacial morphology was observed in the samples subjected to the solid-state aging. Analytical results indicated the electroless Co(W,P) is in essential a combined-type, *i.e.*, sacrificial-type plus stuffed-type, diffusion barrier. However, the $\alpha$-Co(W,P) is a better diffusion barrier for under bump metallurgy (UBM) applications in flip-chip (FC) bonding since it exhibits a lower Co consumption rate in comparison with poly-Co(W,P).

## INTRODUCTION

As a popular interconnection method for advanced electronic packaging, the FC bonding connects the IC chips directly to the substrate in a face-down manner. The bump body and UBM are two essential part of FC bonding in which UBM is a multi-layered structure including adhesion layer, diffusion barrier layer and protective layer. The diffusion barrier layer is to prohibit the interdiffusion between bump body and bond pad of IC to ensure a reliable IC operation. According to mechanism of diffusion retardation, diffusion barrier is classified as sacrificial type, stuffed type, passive-compound type and amorphous type[1]. Electroless plating becomes a competitive process for the preparation of diffusion barriers since it offers the advantages such as low cost, high throughput, good step coverage, *etc.* Uang reported that the stress status of electroless nickel (EN) is lower than that of sputtered Ni[2]. Amorphous EN is showed to be a better diffusion barrier in comparison with sputtered Ni[3-6]. O'Sullivan *et al.* demonstrated that electroless cobalt-phosphorous (Co(P)) possesses good barrier capability in between Cu and interlayer dielectric[7]. Kohn *et al.* showed that addition of W element in electroless Co(P) improves its thermal stability and barrier capability to Cu metallization[8-9]. Studies relating to SAC and EN layers could be found elsewhere[10-17]; however, studies on SAC and electroless Co are relatively less. We hence investigate the barrier characteristics of electroless Co(W,P) to lead-free SAC solder, presently the most popular lead-free solder, so as to explore the applicability of electroless Co(W,P) to UBM in flip-chip Cu-ICs.

## EXPERIMENT

Ti (50 nm)/Cu (100 nm)/Si was chosen as substrate in which the Cu layer is to simulate the Cu interconnects. After the pretreatments depicted in Table 1, about 6-μm thick electroless $\alpha$-Co(W,P) or poly-Co(W,P) layer was deposited on Cu/Ti/Si substrates by adjusting the pH value of plating bath (composition is listed in Table 2) within KOH solution. Eutectic SAC solder paste was applied on the Co(W,P) immediately after electroless plating and a brief reflow at 250°C for 30 sec was carried out. The samples were then vacuum-sealed and sent to furnaces for liquid-state aging at 250°C for 1 hr or for solid-state aging at 150°C up to 1000 hrs. Scanning electron microscopy (SEM, JSM-6500F) and transmission electron microscopy (TEM, Philips Tecnai F-20) in conjunction with the energy dispersive spectroscopy (EDS, Genesis) were adopted to examine the evolutions of microstructure and compositions in the specimens. The cross-sectional TEM (XTEM) samples were prepared by using the focused-ion-beam (FIB, FEI-201) technique supported by Materials Analysis Technology, Inc. at Chupei, Taiwan, R.O.C.

Table1. Chemicals and duration of pretreatments.

| Step | Component | Concentration | Immersion Time |
|---|---|---|---|
| Roughening | $H_2SO_4$ | 5 *wt.%* | 30 sec |
| Sensitization | $SnCl_2 . 2H_2O$ | 10 g/L | 5 min |
| | HCl | 40 ml/L | |
| Activation | $PdCl_2 . 2H_2O$ | 10 g/L | 1 min |
| | HCl | 40 ml/L | |

Table2. Composition of electroless plating bath.

| Component | Concentration ( g/L ) |
|---|---|
| $CoSO_4 . 7H_2O$ | 23 |
| $NaH_2PO_2 . H_2O$ | 18 |
| $Na_3$ Citrate | 144 |
| $H_3BO_3$ | 31 |
| $Na_2WO_4 . 2H_2O$ | 10 |
| pH | 8.6 (for $\alpha$–Co(W,P)) or 7.6 (for poly-Co(W,P)) |
| Temperature | 90°C |

## RESULTS AND DISCUSSION

When SAC solder reacted with Co(W,P) layer, CoSn$_3$ and (Co,Cu)Sn$_3$ IMCs formed at the SAC/Co(W,P) interface. Besides, Ag$_3$Sn IMC was observed in the interior of solder. As shown in Fig. 1(a), the IMCs originally resides at SAC/$\alpha$-Co(W,P) interface spalled into solder after solid-state aging. Furthermore, an about 1.2 μm-thick reaction layer containing Co, Sn, and P elements with P contents as high as 20 *at.%* can be seen in between solder and Co(W,P). It is resulted from the accumulation of P elements when Sn reacts with Co. Above results imply that the $\alpha$-Co(W,P) acts as sacrificial-type barrier to inhibit Sn diffusion. The corresponding EDS line scan profiles are shown in Fig. 1(b), indicating that Co may also stop the interdiffusion between Sn and Cu. Figure 1(c) reveals the XTEM micrograph corresponding to Fig. 1(a). According to the selected area electron diffraction (SAED) attached to the upper right-hand side of Fig. 1(c), the P-rich layer is polycrystalline and comprises of complicated IMCs. Further, the $\alpha$-Co(W,P) becomes polycrystalline and Co$_2$P phase form in Co(W,P) as indicated by the SAED pattern attached at the lower right-hand side of Fig. 1(c). Since Fig. 1(b) indicates negligible Cu diffusion into Co(W,P), it is believed that the Co$_2$P phase might segregate into the grain boundaries of Co(W,P) and thus provide the stuffed-type barrier capability for Co(W,P).

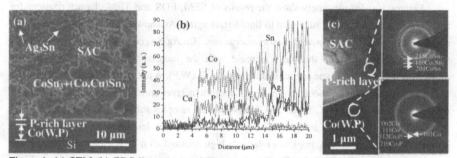

Figure 1. (a) SEM (b) EDS line scan profiles and (c) XTEM views of SAC/$\alpha$-Co(W,P) samples subjected to liquid-state aging.

Figures 2(a)-(c) show the analytical results of SEM, EDS and TEM of the specimen containing SAC/$\alpha$-Co(W,P) subjected to solid-state aging test. As depicted by Fig. 2(a), a thick IMC layer containing CoSn$_3$, (Co,Cu)Sn$_3$ and (Co,Ag)Sn$_3$ resides at the SAC/Co(W,P) interface without spalling into the solder. SEM and TEM analyses both revealed the presence of P-rich layer and, as indicated by SAED pattern attached in Fig. 2(c), it is polycrystalline and contains complicated IMC types. The $\alpha$-Co(W,P) is hence a sacrificial-type barrier to inhibit Sn diffusion, a result similar to that presented in the case of liquid-state aging. TEM analysis also found that the $\alpha$-Co(W,P) remained amorphous after solid-state aging. In conjunction with the EDS line scan profiles shown in Fig. 2(b), the $\alpha$-Co(W,P) layer thus serves as an amorphous-type barrier to inhibit the Cu diffusion during solid-state aging. In conclusion, the $\alpha$-Co(W,P) is a combined-type diffusion barrier, *i.e.*, sacrificial- plus stuffed-type in the case of liquid-state aging and sacrificial- plus amorphous-type in the case of solid-state aging.

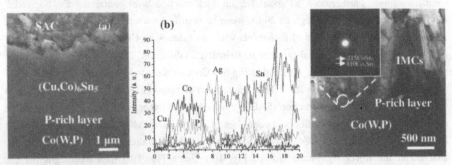

Figure 2. (a) SEM (b) EDS line scan profiles and (c) XTEM views of SAC/$\alpha$-Co(W,P) samples subjected to solid-state aging.

Figures 3(a)-(c) separately show the results of SEM, EDS and TEM characterizations for SAC/poly-Co(W,P) samples subjected to liquid-state aging. As shown in Fig. 3(a), the IMCs are mainly CoSn$_3$ and (Co,Cu)Sn$_3$ and, in some occasions, (Co,Ag)Sn$_3$ could be found. There was no spallation of IMC and its thickness increase with the increase of reflowing time, clearly illustrating the sacrificial-type barrier feature of poly-Co(W,P). A notable feature is shown in Fig. 3(c) which depicts that an about 1.3-$\mu$m thick, amorphous layer enriched with W element ($\sim$ 14.9 *wt.%* W) in beneath the IMC in the SAC/poly-Co(W,P) samples subjected to liquid-state aging, as revealed by TEM/EDS analysis. This is attributed to the less amount of P element in poly-Co(W,P) which, in turn, implies a relatively high accumulation of W element at the reacting interface of solder and Co(W,P). Finally, the SAED analysis indicated the Co$_2$P phase likely presents in the poly-Co(W,P) region and it might offer the stuffed-type barrier capability for poly-Co(W,P) when presenting at the grain boundaries. Hence, the poly-Co(W,P) layer is also a combined-type, *i.e.*, the sacrificial- plus stuffed-type, diffusion barrier layer.

Figures 4(a)-(c) are the analytical results of SEM, EDS and TEM for the specimen containing SAC/poly-Co(W,P) layer subjected to solid-state aging. The IMC at SAC/Co(W,P) interface looks smoother and its thickness is relatively less than that in the sample subjected to liquid-state aging. As shown in Fig. 4(c), TEM similarly revealed an about 300-nm thick, amorphous W-rich layer in between IMC and Co(W,P). Regardless of the aging test type, the arrangement of phases and their constitution presenting in the samples containing poly-Co(W,P) are quite similar and hence, the poly-Co(W,P) is a combined-type, *i.e.*, sacrificial- plus stuffed-type, diffusion barrier.

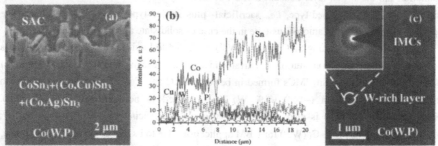

Figure 3. (a) SEM (b) EDS line scan profiles and (c) XTEM views of SAC/poly-Co(W,P) samples subjected to liquid-state aging.

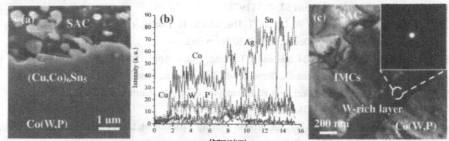

Figure 4. (a) SEM (b) EDS line scan profiles and (c) XTEM views of SAC/poly-Co(W,P) samples subjected to solid-state aging.

Finally, the Co consumption was found to be lower in the case of $\alpha$-Co(W,P). For instance, about 0.6 μm-thick Co was consumed in $\alpha$-Co(W,P) while about 0.9 μm-thick Co was consumed in poly-Co(W,P) after 250°C/1-hr liquid-state aging. In conjunction with the fact of IMC spallation, the $\alpha$-Co(W,P) is expected to be a better diffusion barrier for UBM applications.

## CONCLUSIONS

Diffusion barrier characteristics of electroless $\alpha$-Co(W,P) and poly-Co(W,P) to lead-free SAC solder were investigated. In the case of SAC/$\alpha$-Co(W,P) sample subjected to liquid-state aging, (Co,Cu)Sn$_3$ IMC spalled into solder and an about 1.2-μm thick P-rich layer (~ 20 *wt.%* P) formed in between SAC and Co(W,P). Further, the $\alpha$-Co(W,P) transforms into polycrystalline embedded with tiny $Co_2P$ precipitates. As to the SAC/$\alpha$-Co(W,P) sample subjected to solder-state aging, a thick $(Cu,Co)_6Sn_5$ IMC formed in between SAC and Co(W,P) without spallation and the P-rich layer beneath IMCs was similarly seen. The analytical results indicate the $\alpha$-Co(W,P) is combined-type, *i.e.*, sacrificial- plus stuffed-type in the case of liquid-state aging and sacrificial- plus amorphous-type in the case of solid-state aging, diffusion barrier.

The samples containing SAC/poly-Co(W,P) subjected to liquid- and solid-state aging tests possessed similar phase constitution and arrangement at the reacting interfaces. The mixture of $(Co,Cu)Sn_3$ and $(Co,Ag)Sn_3$ IMCs formed in between SAC and Co(W,P) and an about 300-nm thick, amorphous W-rich (~14.9 *wt.%* W) layer formed in between IMC and Co(W,P). Electroless poly-Co(W,P) is also a combined-type, *i.e.*, sacrificial- plus stuffed-type barrier; however, it is inferior to $\alpha$-Co(W,P) for UBM applications due to its higher Co consumption rate.

## REFERENCES

1. M. A. Nicolet, Thin Solid Films **52**, 415 (1978).
2. R. H. Uang, K. C. Chen, S. W. Lu, H. T. Hu, and S. H. Huang, (IEEE Electron. Packaging. Technol. Conf., Singapore, December 5-7 2000), p.292.
3. M.W. Liang, T.E. Hsieh, C.C. Chen and Y.T. Hung, Jpn. J. Appl. Phys. **43**, 8258 (2004).
4. T.Oppert, E. Zakel, and T. Teutsch, (Proc. IEMT/IMC Symp., Japan, April15-17 1998), p.106.
5. T. Teutsch, T. Oppert, E. Zakel, and E. Klusmann, (Electron. Comp. and Technol. Conf., Las Vegas, NV, Piscataway, NJ USA, 2000), p.107.
6. G. O. Mallory and J. B. Hajdu, Electroless Plating Fundamentals and Applications, (AESF Orlando, Florida 1990) Chap. 1-7.
7. E. J. O'Sullivan, A. G. Schrott, M. Paunovic, C. J. Sambucetti, J. R. Marino, P. J. Bailey, S. Kaja, and K. W. Semkow, IBM J. Res. and Develop. **42**, 607 (1998).
8. A. Kohn, M. Eizenberg, Y. Shacham-Diamand, and Y. Sverdlov, Mater. Sci. Eng. **A302**, 18 (2001).
9. A. Kohn, M. Eizenberg, Y. Shacham-Diamand, and Y. Sverdlov, Microelectron. Eng. **55**, 297 (2001).
10. C. E. Ho, R. Y. Tsai, Y. L. Lin, and C. R. Kao, J. Elect. Mater. **31**(6), 584 (2002).
11. Chi-Won Hwang and Katsuaki Suganuma, J. Mater. Res. **18**(11), 2540 (2003).
12. Yung-Chi Lin and Jenq-Gong Duh, Scripta Mater. **54**, 1661 (2006).
13. Yung-Chi Lin and Jenq-Gong Duh, Scripta Mater. **56**, 49 (2007).
14. V. Vuorinen, T. Laurila, H. Yu, and J. K. Kivilahti, J. Appl. Phys. **99**, 023530 (2006).
15. Yung-Chi Lin, Jenq-Gong Duh, and Bi-Shiou Chiou, J. Elect. Mater. **35**(1), 9 (2006).
16. Yung-Chi Lin, Jenq-Gong Duh, and Bi-Shiou Chiou, J. Elect. Mater. **35**(8), 1665 (2006).
17. Yung-Chi Lin, Jenq-Gong Duh, and Bi-Shiou Chiou, J. Elect. Mater. **36**(11), 1469 (2007).

# Metallization I

Mater. Res. Soc. Symp. Proc. Vol. 1156 © 2009 Materials Research Society       1156-D04-02

## Atomic Layer Deposition of Ruthenium Films on Hydrogen Terminated Silicon

S. K. Park,[1] K. Roodenko,[1] Y. J. Chabal,[1] L. Wielunski,[2] R. Kanjolia,[3] J. Anthis,[3] R. Odedra,[3] and N. Boag[4]

[1] Materials Science and Engineering Department, University of Texas at Dallas,
Richardson, Texas 75080, USA
[2] Laboratory for Surface Modification, Rutgers University,
Piscataway, NJ 08854, USA
[3] SAFC Hitech,
Haverhill, MA 01832, USA
[4] Functional Materials, Institute for Materials Research, University of Salford,
Salford, Manchester M5 4WT, UK

## ABSTRACT

Atomic Layer deposition of thin Ruthenium films has been studied using a newly synthesized precursor (Cyclopentadienyl ethylruthenium dicarbonyl) and $O_2$ as reactant gases. Under our experimental conditions, the film comprises both Ru and $RuO_2$. The initial growth is dominated by Ru metal. As the number of cycles is increased, $RuO_2$ appears. From infrared broadband absorption measurements, the transition from isolated, nucleated film to a continuous, conducting film (characterized by Drude absorption) can be determined. Optical simulations based on an effective-medium approach are implemented to simulate the *in-situ* broadband infrared absorption. A Lorentz oscillator model is developed, together with a Drude term for the metallic component, to describe optical properties of Ru/$RuO_2$ growth.

## INTRODUCTION

Ruthenium and ruthenium oxides exhibit high chemical and thermal stabilities in presence of high-κ dielectric materials such as hafnium oxide and aluminum oxide.[1] Additionally they have relatively high work functions, which is important for gate metals.[2] Thus, they are expected to provide a barrier and seed layer for copper deposition, an important interconnection material in microelectronics, a capacitor electrode material for memory devices, and a gate metal for metal-oxide-semiconductor field effect transistors.[1, 3-8] There has therefore been an increased effort to study the deposition of Ru and $RuO_2$ using different growth techniques.[2, 9-11] Since atomic layer deposition is a powerful method to deposit thin, uniform, and conformal films even on structured surfaces such as trenches and via holes, there is an active search for ruthenium ALD precursors possessing appropriate physical and chemical properties for the deposition process.

In this study, we present *in-situ* Fourier transform infrared (FTIR) studies of ALD growth of ruthenium thin films on hydrogen-terminated silicon (111) surfaces using cyclopentadienyl ethylruthenium dicarbonyl precursor [Ru(Cp)(CO)$_2$Et], one of the newly synthesized cyclopentadienyl dicarbonyl alkyl ruthenium precursors, with $O_2$ as oxidant. The broadband infrared absorption obtained from in-situ FTIR measurements is simulated using optical models, in order to obtain an idea of evolution of the Ru, $RuO_2$ and $SiO_2$ species on the surface.

**EXPERIMENTAL PROCEDURES AND MODELING**

For this study, double-sided polished, Si (111) wafers are passivated with hydrogen by an immersion in HF and NH₄F after a standard RCA clean, followed by deionized water rinse and a N₂ gas drying cycle.[12] Growth is performed in a home-built ALD reactor connected to a FTIR spectrometer.[13] In this system, deposition is performed using cyclopentadienyl ethylruthenium dicarbonyl [Ru(Cp)(CO)₂Et] and O₂ as reactant gas. The IR beam from the spectrometer is focused on the sample at the Brewster angle (74°). The transmitted IR beam is refocused onto a liquid nitrogen-cooled Mercury Cadmium Telluride (MCT-B) detector.

The infrared absorption spectra are fitted using the dielectric function in the form of a Lorenzian oscillator model with the Drude term for conductivity (eq.1)

(eq. 1) $\varepsilon(w) = \varepsilon_\infty - \frac{w_N^2}{w^2 + i\Gamma_D w} + \sum_k f_k \frac{w_{0k}^2}{w_{0k}^2 - w^2 + i\Gamma_k w}$

where $\varepsilon_\infty$ is a high-frequency dielectric function, $w_N^2 = \varepsilon_\infty\, w_p^2$ with $w_p$ the plasma frequency (in eV), $\Gamma_D$ is the Drude-term damping factor (in eV), $f_k$ is the Lorenz-term oscillator strength, $w_{0k}$ is the resonance frequency of the k$^{th}$ oscillator, and $\Gamma_k$ is the related damping factor.[14]

To simulate the dielectric response of the mixed layers, we used the effective medium approximation (EMA) using the Maxwell-Garnet form:[15]

(eq.2) $\varepsilon_{eff}(w) = \varepsilon_B \frac{1 + 2f_A[(\varepsilon_A - \varepsilon_B)/(\varepsilon_A + 2\varepsilon_B)]}{1 - f_A[(\varepsilon_A - \varepsilon_B)/(\varepsilon_A + 2\varepsilon_B)]}$

where $\varepsilon_A$, $\varepsilon_B$ are the dielectric functions of the layer composites, and $f_A$ is the fraction of composite A in the layer.

For Ru and RuO₂ composites, we follow the use of the Drude and Lorentz terms as proposed by *Hones*[14] *and Choi*[16] to separate the free electrons (described by Drude term) from the bound charge (optical transitions described by Lorentz term). In the calculations we used the following high-frequency dielectric functions: $\varepsilon_\infty$=2.82 for RuO₂[14], and $\varepsilon_\infty$=2.0 for Ru[16]. The Lorenz parameters were fixed in our fits at the values given by *Hones*[14] *and Choi*,[16] except for the lower-energy parameter that has a big effect on the IR spectral range in Ru. *Choi et al.*[16] report this lower-energy parameter to be at 0.6 eV.[16] We allowed this parameter to vary within 15% of the literature value, as well as the related damping constant and the oscillator strength. Such changes in optical constants may be related to the dynamic changes in layer homogeneity and quality with the growth of the Ru layer and its interaction with the Si substrate and the interfacial layer. Also the Drude term $w_N$ was allowed to vary. The best overall value for RuO₂ was 3±1 eV, with an increase to 5±1 eV after 24 cycles. This value is close to the value of 5.22 eV which was reported by *Hones et al.*[14] for the bulk RuO₂. For Ru, *Choi et al.*[16] do not report on Drude-term parameters, but from their plots a plasma frequency of $w_N$=7 eV is estimated with a damping constant of 0.3 eV. In our fits, the plasma frequency fluctuates with the lowest value of $w_N$=5.5 eV. After 24 cycles, the highest value of $w_N$=7.5±0.5eV is obtained, which is close to the value extracted from ref. 16. This suggests that the free electron contribution becomes higher as more Ru is deposited.

To simulate radiation propagation within the deposited film, we used a 2x2 matrix formalism for isotropic structures.[17] Since the measured data did not show any interference fringes, an incoherent-type model was applied in accordance to ref.[18,19]. The schematic drawing of the model is shown in Fig. 1. The model includes Ru on both sides of the Si substrate, in a mirrored version of Fig. 1.

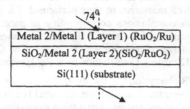

| 74 |
|:---:|
| Metal 2/Metal 1 (Layer 1) (RuO₂/Ru) |
| SiO₂/Metal 2 (Layer 2)(SiO₂/RuO₂) |
| Si(111) (substrate) |

Fig.1. Layer model which was used to simulate our IR spectra.

In the current model, we did not include the fit on the $SiO_2$ absorption bands, fitting only on the base-line of the measured IR absorbance. However, we checked that the inclusion of the absorbance due to i. e. Si-O vibrations does not have a strong influence on the baseline.

## RESULTS AND DISCUSSION

The chemistry of Ru ALD will be reported elsewhere.[20] Briefly, the Ru precursor is observed to react with an atomically flat H/Si (111) at 300°C. The Si-H stretching mode [2083 cm⁻¹] redshifts to 2059 cm⁻¹ and decreases in intensity. Only the hydrocarbon-containing ligand [C-H peaks from the ethyl group at ~2900 cm⁻¹] is observed (CO and Cp not detected). Upon exposing the surface to $O_2$ at 300°C, this ethyl group attached during the first 0.5 cycle is removed, as expected for ALD growth. This is consistent with the observation that, in the Ru ALD process, $O_2$ oxidatively decomposes the ligands of the metal precursors.[21] Once Ru is deposited, oxygen dissociatively adsorbs on the noble metal surface, which subsequently promotes the reaction of oxygen with the ligand.[22, 23]

*Ex-situ* RBS measurements are particularly useful to determine the absolute coverage of Ru atoms, and therefore to evaluate the average thickness of the ALD grown films, making simple assumptions about the Ru atom densities (Fig. 2). For the first two cycles, the growth rate is low. However, beyond that number, the growth rate increases more rapidly and is linear after 3 cycles.

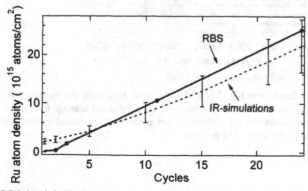

Fig. 2. RBS (straight line) and the data from IR simulations (broken line) are shown.

To determine the chemical states and evaluate whether the films are metallic Ruthenium or Ruthenium oxide, *ex-situ* XPS measurement were performed. For the sample deposited with 12 cycles and 24 cycles, two contributions corresponding to pure Ru and to Ru oxide are observed. While the relative intensity of Ru oxide component increases with the number of ALD cycles, there is still a co-existence of both Ru and Ru oxide. The O1s spectra show that Ru dominates initially (e.g. after 12 cycles). Contribution from $RuO_2$ increases after 24 cycles.

In the analysis of the IR absorption spectra, the focus is on the broadband absorption. All spectra are referenced to the spectrum obtained from the initial H/Si (111) surface. Fig. 3 shows the IR spectra along with the simulated results. From these simulations, the "nominal thickness" can be extracted as a function of ALD cycles. These thickness, $d_{Ru}$ and $d_{RuO2}$, are defined as the fraction of the material in EMA model multiplied by the thickness of the simulated layers, as shown in Fig. 4. Our fit appears to underestimate $SiO_2$ amount, but that may be related to the fact that our model includes a mixed $SiO_2/RuO_2$ layer only, while the sample may include also $SiO_2/Si$ phases. Including additional layers would certainly improve the fit, but at the cost of additional parameters which we wanted to avoid at this stage. From the nominal thicknesses, the Ru atom density can be extracted using eq. 3:

$$\text{(eq. 3)} \quad \rho_{total} = \rho_{Ru} d_{Ru} + \rho_{RuO2} d_{RuO2}$$

where $\rho_{total}$ is the total Ru density (see Fig. 2), and $\rho_{Ru}$ and $\rho_{RuO2}$ the Ru and $RuO_2$ densities, $\rho_{Ru}$ and $\rho_{RuO2}$ are $7.27 \times 10^{22}$ atoms/cm$^3$ for bulk Ru and $3.15 \times 10^{22}$ atoms/cm$^3$ for bulk $RuO_2$.

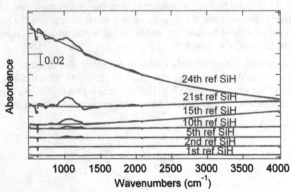

Fig. 3. Measured IR data (black) and simulations (red).

The data shown by a broken line in Fig. 2 is within the error bars from the measured RBS data for the higher cycles. For the initial cycles however the model did not manage to reproduce well the RBS data. This may be due to the different layer system in the initial cycles, where for example a mixed air/Ru/Si phase could be present.

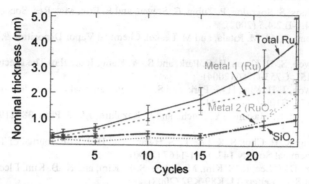

Fig. 4. Evolution for Ru, RuO$_2$, and SiO$_2$ species as obtained from the IR simulations. "Total Ru" is the sum of Ru and RuO$_2$ thicknesses.

## CONCLUSIONS

The growth of Ru/RuO$_2$ films using ALD was performed with a newly synthesized precursor Ru(Cp)(CO)$_2$Et and O$_2$. Beyond an incubation period (first two cycles), the growth rate is linear. Under our experimental conditions, the film comprises both Ru (initially dominant) and RuO$_2$ (apparent beyond 12 cycles). After ~22 cycles, the broadband IR absorption changes qualitatively, displaying a relatively stronger absorption at *lower* wavenumbers. This behavior is characteristic of free electron absorption (Drude absorption) and indicates that conductive grains coalesce to form a continuous film. The optical effective-medium model used in our fits included two layers, of Ru/RuO$_2$ and RuO$_2$/SiO$_2$ mixed phases. Without depth profiling, it is always complicated to choose the appropriate layer structure, but our results show a reasonable agreement with the amount of Ru and RuO$_2$ estimated from RBS and XPS measurements, which also show the coexistence SiO$_2$, RuO$_2$ and Ru phases when RuO$_2$ is formed.

## ACKNOWLEDGMENTS

Funding for this work was provided by SAFC Hitech and NSF (CHE-0827634). We thank Paul Gevers from Plasma & Materials Processing Group at Eindhoven University of Technology for useful discussions and literature summary on Ru and RuO$_2$ optical constants.

1   F. Papadatos, S. Spyridon, P. Zubin, C. Steven, and E. Eric, Mat. Res. Soc. Symp. Proc. **716**, B2.4.1-B.2.4.5 (2002).

2   T. Aaltonen, P. Al, M. Ritala, and M. Leskel, Chemical Vapor Deposition **9**, 45-49 (2003).

3   O.-K. Kwon, S.-H. Kwon, H.-S. Park, and S.-W. Kang, Journal of The Electrochemical Society **151**, C753-C756 (2004).

4   O.-K. Kwon, J.-H. Kim, H.-S. Park, and S.-W. Kang, Journal of The Electrochemical Society **151**, G109-G112 (2004).

5   A. Tomonori, K. Masahiro, Y. Soichi, and E. Kazuhiro, Jpn. J. Phys. **38**, 2194-2199 (1999).

6   S. Y. Kang, K. H. Choi, S. K. Lee, C. S. Hwang, and H. J. Kim, Journal of The Electrochemical Society **147**, 1161-1167 (2000).

7   S.-S. Yim, D.-J. Lee, K.-S. Kim, M.-S. Lee, S.-H. Kim, and K.-B. Kim, Electrochemical and Solid-State Letters **11**, K89-K92 (2008).

8   S. Y. Kang, C. S. Hwang, and H. J. Kim, Journal of The Electrochemical Society **152**, C15-C19 (2005).

9   T. Aaltonen, M. Ritala, K. Arstila, J. Keinonen, and M. Leskel, Chemical Vapor Deposition **10**, 215-219 (2004).

10  Y.-H. Lai, Y.-L. Chen, Y. Chi, C.-S. Liu, A. J. Carty, S.-M. Peng, and G.-H. Lee, Journal of Materials Chemistry **13**, 1999-2006 (2003).

11  S.-E. Park, H.-M. Kim, K.-B. Kim, and S.-H. Min, Thin Solid Films **341**, 52-54 (1999).

12  G. S. Higashi, Y. J. Chabal, G. W. Trucks, and K. Raghavachari, Appl. Phys. Lett. **56**, 656 (1990).

13  Y. Wang, M. Dai, M.-T. Ho, L. S. Wielunski, and Y. J. Chabal, Appl. Phys. Lett. **90**, 022906 (2007).

14  P. Hones, T. Gerfin, and M. Gratzel, Applied Physics Letters **67**, 3078-3080 (1995).

15  S. Norrman, T. Andersson, C. G. Granqvist, and O. Hunderi, Physical Review B **18**, 674 (1978).

16  W. S. Choi, S. S. A. Seo, K. W. Kim, T. W. Noh, M. Y. Kim, and S. Shin, Physical Review B (Condensed Matter and Materials Physics) **74**, 205117-8 (2006).

17  N. M. Bashara and R. M. A. Azzam, in *Ellipsometry and Polarized Light*, 3rd ed. (Elsevier Science Amsterdam, 1989).

18  R. F. Potter, in *Handbook of optical solids* (Academic Press San Diego, 1985).

19  B. Harbecke, Appl. Phys. B. **39**, 165 (1986).

20  S. K. Park, K. Roodenko, Y. J. Chabal, R. Kanjolia, J. Anthis, R. Odedra, and N. Boag, (Unpublished).

21  T. Aaltonen, A. Rahtu, M. Ritala, and M. Leskela, Electrochemical and Solid-State Letters **6**, C130-C133 (2003).

22  S. K. Kim, S. Y. Lee, S. W. Lee, G. W. Hwang, C. S. Hwang, J. W. Lee, and J. Jeong, Journal of The Electrochemical Society **154**, D95-D101 (2007).

23  Y. Matsui, M. Hiratani, T. Nabatame, Y. Shimamoto, and S. Kimura, Electrochemical and Solid-State Letters **4**, C9-C12 (2001).

Mater. Res. Soc. Symp. Proc. Vol. 1156 © 2009 Materials Research Society

## Coupled Finite Element – Potts Model Simulations of Grain Growth in Copper Interconnects

Bala Radhakrishnan[1] and Gorti Sarma[1]
[1]Computer Science and Mathematics Division, Oak Ridge National Laboratory,
Oak Ridge, TN 37831-6164

## ABSTRACT

The paper addresses grain growth in copper interconnects in the presence of thermal expansion mismatch stresses. The evolution of grain structure and texture in copper in the simultaneous presence of two driving forces, curvature and elastic stored energy difference, is modeled by using a hybrid Potts model simulation approach. The elastic stored energy is calculated by using the commercial finite element code ABAQUS, where the effect of elastic anisotropy on the thermal mismatch stress and strain distribution within a polycrystalline grain structure is modeled through a user material (UMAT) interface. Parametric studies on the effect of trench width and the height of the overburden were carried out. The results show that the grain structure and texture evolution are significantly altered by the presence of elastic strain energy.

## INTRODUCTION

In copper interconnects, the grain structure and texture in the copper film evolve both by the process of self-annealing at room temperature and during a subsequent elevated temperature anneal. In self-annealing the elastic stress and strain in the film are tensile as the film contracts more than the surrounding dielectric on cooling from the deposition temperature, whereas during an elevated temperature anneal, the stress and strain in the film are compressive. In the presence of thermal expansion mismatch, the elastic anisotropy of copper leads to significant variations in the elastic stored energy in the copper grains as a function of the grain orientation. Grain boundary migration is therefore governed by two driving forces–grain boundary curvature and the elastic stored energy difference across the boundaries. The grain size and the grain boundary character distribution of the copper grains are significant parameters because they control both the electrical resistance and the resistance to electro-migration led failures in the copper lines. Grain growth and texture in copper interconnects have been extensively studied both experimentally and through modeling [1-6]. However, there is no clear consensus on the influence of trench and overburden geometry on the final grain morphology and texture. The present study represents an early attempt in simulating grain structure evolution at the mesoscale driven by anisotropic stored energy differences and curvature in the presence of anisotropic grain boundary energy and mobility. Parametric studies were carried out by systematically varying the trench aspect ratio and the thickness of the overburden for two levels of thermal expansion mismatch.

## THEORY

The grain growth and texture evolution in the copper film are significantly influenced by the initial texture in the as-deposited film because the grain boundary energy and mobility that govern the grain boundary migration depend on the local boundary misorientation. Therefore, it is important to start with an initial grain structure with realistic grain orientations. The initial grain orientations were assigned based on the experimental work by Besser et al. [1]. Typically, there are three sets of grain orientations – a {111}<uvw> component that nucleates from the trench bottom and the trench sides, and a weaker {511}<uvw> component that occurs due to twinning. In the current simulations, the sidewall {111}<uvw> nucleation was limited to a fixed distance from the sidewall, so that its strength decreased with increasing trench width. The texture assignments also satisfied the experimental observation that the {111}<uvw> orientations that nucleated from the bottom of the trench also had an in-plane preference of <aa0> directions. In the overburden, a fixed fraction of the grains was assigned near-Cube orientations and the remaining grains were assumed to be of the {111}<uvw> type. Typical (111) pole figures obtained from the work of Besser et al. [1] and a typical initial texture used in the simulations are shown in figure 1.

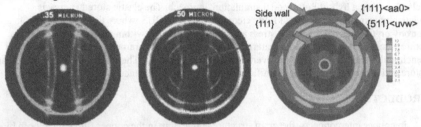

**Figure 1**. (111) pole figures showing experimental textures from the work of Besser et al. [1] (left and middle) and typical starting texture used in the present simulations (right).

The first step in the computational approach is the finite element modeling of the copper and the dielectric to calculate the magnitude and spatial distribution of the elastic stored energy density in the copper grains when subjected to a high temperature anneal. Figure 2 shows a typical geometry used in the analysis to model the thermal expansion mismatch between copper and the dielectric, which in this case is made up of silica on the sides and silicon nitride at the bottom. The boundary conditions in the $X$ and $Z$ directions are as shown in figure 2. The displacement in the $Y$ direction was assumed to be zero in order to simulate plane strain. Two levels of thermal expansion mismatch corresponding to annealing temperatures of (200+ T)°C and (400+T)°C were investigated, where T is the zero-stress temperature, which was assumed to be room temperature. The elastic properties were assumed to be isotropic in silica and silicon nitride and anisotropic in copper. Simulations were carried out using the commercial finite element code ABAQUS with a user material subroutine (UMAT) to specify the anisotropic properties for copper grains. The copper grains were discretized such that there were many elements within each grain. The elasticity tensor for each element in the global system was obtained by rotation of the local tensor based on the grain orientation. The UMAT was also used to calculate the stress and strain energy density in each element.

The next step was to map the strain energy density and the element orientations to a square grid in order to perform the grain growth simulations using the hybrid Potts model. Potts model has been used extensively for simulating curvature driven grain growth in polycrystals [7]. It accurately captures the grain growth exponent, grain topology and the grain size distribution associated with curvature-driven growth. However, the probabilistic site-flipping rules used in

Potts model to simulate curvature driven growth are not consistent with site-flipping rules required to accurately capture grain boundary migration driven by a bulk stored energy difference across the grain boundary. One approach to over come this problem is to use a hybrid model [7], where a separate set of flipping rules for curvature driven and strain energy driven migration are used with a relative

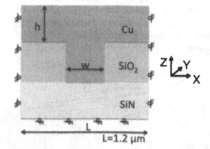

| Material | CTE [/°C] | Young's modulus [GPa] | Poisson's ratio |
|----------|-----------|------------------------|-----------------|
| Cu | $16.6 \times 10^{-6}$ | 110.3 | 0.34 |
| $SiO_2$ | $0.55 \times 10^{-6}$ | 73.0 | 0.17 |
| SiN | $3.2 \times 10^{-6}$ | 220.8 | 0.27 |

**Figure 2**. Geometry of the copper trench and the dielectrics and the corresponding material properties used in the finite element modeling of thermal mismatch strains.

frequency that depends on the magnitude of the two driving forces. The above approach was used in the current simulations. The simulations were used to carry out a parametric study of the effect of trench width, overburden thickness and the level of mismatch strain on the grain growth kinetics and texture evolution in copper. The various simulation cases are shown in Table I. Cases 1a to 6a correspond to the same geometries as in cases 1 to 6 with a higher annealing temperature difference (400°C).

**Table I**. Simulation parameters used in the current study

| Case # | Trench width (µm) | Overburden height (µm) | Temperature difference (deg C) |
|--------|-------------------|------------------------|--------------------------------|
| 1 | 0.6 | 0.15 | 200 |
| 2 | 0.6 | 0.075 | 200 |
| 3 | 0.6 | 0.225 | 200 |
| 4 | 0.225 | 0.225 | 200 |
| 5 | 0.15 | 0.225 | 200 |
| 6 | 0.10 | 0.225 | 200 |

## DISCUSSION

Figure 3a shows the mean grain size as a function of the simulation time. The size of each site used in the Potts model simulations was $2 \times 10^{-9}$ m. Therefore the grain size numbers shown in figure 3a have to be multiplied by $2 \times 10^{-9}$ to obtain the actual grain size. The simulation time is shown in Monte Carlo Step (MCS) which is the time corresponding to one flip attempt per site in the domain. It can be shown that the real time is related to MCS through a jump frequency, the real temperature and the activation energy for grain boundary migration [8]. It can be seen that

81

all simulations that were carried out by including the elastic strain energy driving force had a higher growth rate than the simulations that considered only the driving force due to curvature. Therefore, it is clear that the elastic stored energy played a significant role in grain growth kinetics. Figure 3b shows the variation of the elastic strain energy in the domain as a function of the simulation time. It is clear that both for the high and the low temperature anneal, the stored energy decreased with time, indicating that grain growth occurred by the preferential growth of low stored energy grains at the expense of the high stored energy grains. However, the shape of the grain growth curve still indicates parabolic kinetics, although simulations carried out to long times showed sudden increase in growth rate associated with abnormal grain growth.

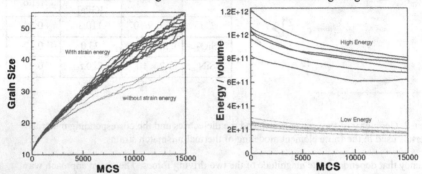

**Figure 3**. (a) Grain size and (b) elastic strain energy as a function of simulation time.

Figures 4 and 5 show the elastic stored energy distribution in the grains before and after grain growth. It is clear that grain growth is accompanied by a reduction in the elastic strain energy. The contour levels used for the energy are the same for the grain structures before and after anneal. It is clear that the high energy sites before annealing are replaced by lower energy orientations after grain growth. For annealing times used in the simulations, there is no clear trend on the effect of overburden on the final grain size in the trench for the low temperature anneal. However, as shown in figure 5 for the high temperature annealing cases, trench width has significant effect on the grain size. The average grain size in the widest trench (case 4a) is significantly higher than in the narrowest trench (case 6a). However, for the simulation times used in this study, the grain growth did not progress far enough to produce a single crystal at the bottom.

Investigation of texture before and after annealing shows that the texture components depend significantly on the trench width and the overburden thickness. In general, the sidewall {111}<uvw> components and the {511}<uvw> components grow at the expense of the bottom wall {111}<uvw> components. Figure 6a shows the (111) pole figures before and after the low temperature anneal as a function of overburden height. An increase in the side wall {111}<uvw> is seen for case 1, while it decreases for case 2 and case 3. In case 2, bottom wall {111}<aa0> components are seen clearly, while in case 3 the {511}<uvw> components are seen to increase at the expense of bottom wall {111}<aa0> components.

Figure 6b shows the (111) pole figures from the experimental work of Besser et al. [1] for the case of a low and a high temperature anneal. It clearly shows the sidewall {111}<uvw> component and the bottom wall {111}<uvw> component with intensities near the {111}<aa0>

locations. The simulations under certain conditions capture these components, while not in other cases. Future work with more accurate starting textures is required to better capture the experimental textures.

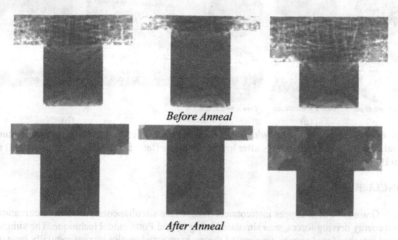

*Before Anneal*

*After Anneal*

**Figure 4**. Elastic strain energy distribution in copper grains, before and after the low temperature anneal. The effect of overburden height is not significant.

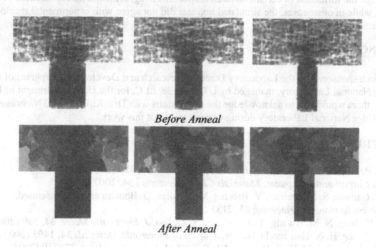

*Before Anneal*

*After Anneal*

**Figure 5**. Elastic strain energy distribution, before and after the high temperature anneal. The grain size in the widest trench is significantly higher than in the narrowest trench.

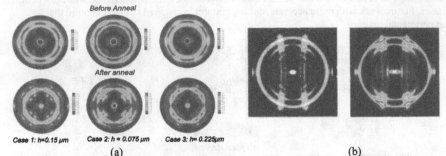

Before Anneal

After anneal

Case 1: h=0.15 μm     Case 2: h = 0.075 μm     Case 3: h= 0.225μm

(a)                                                         (b)

**Figure 6**. (111) pole figures showing (a) simulated textures before and after low temperature anneal, and (b) measured textures after low temperature (left) and high temperature (right) anneal from Besser et al. [1].

## CONCLUSIONS

Grain growth in copper interconnect lines, in the simultaneous presence of curvature and stored energy driving forces, was simulated using hybrid Potts model technique. The simulations showed that the stored energy accelerated the grain growth kinetics by preferentially growing the low energy orientations at the expense of the high-energy orientations. The resulting textures were very sensitive to the trench geometry and the overburden thickness. The simulations predicted the formation of certain texture components in agreement with experiments in certain cases, while in other cases, the simulated textures did not agree with experimental results of Besser et al. [1].

## ACKNOWLEDGMENTS

Research sponsored by the Laboratory Directed Research and Development Program of Oak Ridge National Laboratory, managed by UT-Battelle, LLC, for the U. S. Department of Energy. The authors would like to acknowledge the discussions with Drs. Kulkarni and Nicholson of the Oak Ridge National Laboratory during various stages of this work.

## REFERENCES

1. P. R. Besser et al., *Journal of Electronic Materials* **30**, 320 (2001).
2. K. Mirpuri and J. Szpunar, *Materials Characterization* **54**, 1007 (2005).
3. V. Carreau, S. Maitrejean, Y. Brechet, M. Verdier, D. Bouchu and G. Passemard, *Microelectronic Engineering* **85**, 2133 (2008).
4. J. K. Jung, N. M. Hwang, Y. J. Park and Y. C. Joo, *J. Electronic Mater.* **34**, 559 (2005).
5. H. J. Lee, H. N. Han and D. N. Lee, *Journal of Electronic Materials* **34**, 1493 (2005).
6. M. O. Bloomfield, D. N. Bentz and T. S. Cale, *J. Electronic Mater.* **37**, 249 (2008).
7. A. D. Rollett and D. Raabe, *Computational Materials Science* **21**, 69 (2001).
8. B. Radhakrishnan and T. Zacharia, *Metall. Mater. Trans A* **26**, 2123 (1995).

Mater. Res. Soc. Symp. Proc. Vol. 1156 © 2009 Materials Research Society　　　1156-D04-08

# Electroless Cu Deposition on Self-Assembled Monolayer Alternative Barriers

S. Armini[a], A. Maestre Caro[a,b]

[a] *IMEC, Kapeldreef 75, B-3001 Heverlee, Leuven, Belgium*
[b] *Katholieke Universiteit Leuven, Kasteelpark Arenberg 1, B-3001 Heverlee, Belgium*

* corresponding author: Silvia Armini, IMEC, Kapeldreef 75, B-3001 Leuven, Belgium. Tel.: +32 (0) 16 28 86 17. Fax. : +32 (0) 16 28 13 15. E-mail: silvia.armini@imec.be

## ABSTRACT

An alternative bottom-up Cu electro-less deposition (ELD) method without other catalyst material activation tested on blanket wafers, is the focus of this paper. The process consists in reducing the Cu ions via standard reducing agents, such as dimethylamine borane (DMAB). A wide range of experimental conditions such as pH, temperature, Cu ion concentration and time are investigated and the Cu layer nucleation and growth mechanism is evaluated on clean $SiO_2$ and after functionalization with 3-aminopropyltrimethoxysilane (APTS) self-assembled monolayer (SAM) used as copper diffusion barrier. The barrier properties of the APTS layer after Cu ELD are also assessed by copper resistivity measurements and visual inspections as a function of the annealing temperature.

## INTRODUCTION

Due to the several applications of metalized organic layers in organic/molecular electronics [1], polymer light emitting diodes [2], food packaging and prosthetic implants [3], there is an increasing interest in understanding and controlling the interaction between metals and organic films [4, 5, 6, 7]

Cu is the interconnect material of choice in the metallization step for advanced semiconductor device manufacturing. A typical interconnect structure is composed of Cu/Ta/TaN/SiO2 where the Ta/TaN layer is required as a barrier to prevent interdiffusion between Cu and the underlying dielectric resulting in electrical shorting. Due to its poor electrical conductivity, relative to Cu, the barrier layer thickness must be minimized while maintaining high performance diffusion barrier properties and good adhesion strength with neighboring layers. Nevertheless, barrier layer formation has become increasingly difficult as the technology node is reduced. An alternative to the conventional barrier process is an organic "self-assembling" system, so-called self-assembled monolayer (SAM). This inherently bottom-up process involves the anchoring of bi-functionalized organic molecules on the inter-layer dielectric (ILD) surface. There are many potential advantages of SAMs vs. metal-based barriers: i) conformality, ii) coverage in high aspect ratio (HAR) structures, iii) selectivity, iv) tailored surface functionalization to control reactivity, v) modulation of line resistance.

The current approaches to Cu metallization include chemical vapor deposition (CVD), physical vapor deposition (PVD), selective electroless deposition (ELD) and electroplating. As device sizes decrease, accommodated by scaling and materials changes, electrochemical deposition is considered the most promising method due to its many advantages such as good uniformity and gap filling ability, selectivity and low processing temperatures. Due to the need for applied power and non-uniform current distribution of Cu electroplating, Cu ELD is especially emphasized for future interconnect technologies. In this method deposition occurs via the chemically promoted reduction of metal ions without an externally applied potential. It is therefore a soft deposition technique expected having the potential to eliminate or greatly reduce the metal penetration through the SAMs observed for medium-to-low-reactivity metals, such as copper [8]. A conventional ELD approach suffers some shortcomings such as the high price of typical metal catalysts (Pt, Pd…) and possible damage to the electric properties of the Cu film due to the presence of catalyst as a contaminant. In particular, the presence of this noble metal layer may increase the resistance of the interconnect lines causing current leakage (e.g. the presence of a catalyst Pd layer may lead to a 20-50% increase in Cu interconnect line resistance) [9]. In addition, pre-deposition seeding steps are normally not spatially selective, although methods such as microcontact printing and scanning probe lithographies allow selective deposition [10].

In this paper, an alternative bottom-up Cu ELD method without other catalyst material activation was developed. It consists in the reduction the Cu ions via standard reducing agents and mild experimental conditions (i.e. temperature, pH, reaction time, and concentrations).

## EXPERIMENT

*SAM deposition.* For the SAM deposition experiments, 2 x 2 cm Si (100) coupons were used as substrates. SAMs derived from the APTS (≥ 97.0 %, Sigma Aldrich) molecules were deposited on ca. 2 nm SiO2 layer formed on the Si surface after a cleaning step consisting in 15 min. exposure in a UV-ozone treatment in a Jelight UVO-Cleaner. The deposition procedure consisted in the immersion of the clean SiO2 surface in a 5 mM APTS solution for 30 min. using toluene as a solvent. To remove any impurity, unreacted silane molecules, and solvent residue the coupons were consecutively rinsed and sonicated for 30 s in clean toluene, acetone and de-ionized water (DIW). All the chemicals were used as supplied. Finally the coupons were dried under nitrogen flow.

*Cu ELD.* Dimethylamine borane (DMAB) and sodium citrate (Cit) have been chosen as a reducing and a complexing agent respectively, in order to be able to control the Cu ELD rate. Only fresh solutions were used for each soaking experiment. A DOE has been carried on in order to optimize the critical deposition parameters such as temperature, time, Cu ion concentration and pH. The optimized ELD bath composition is summarized in Table I. After copper ELD, the samples were rinsed thoroughly with DI water and dried under nitrogen gas.

*Contact angle (CA) measurements.* The macroscopic order after the deposition step was routinely monitored measuring the water CA on the SAMs surface. CA measurements were performed using an OCA 20 video-based device from Dataphysics. Measurements were made on sessile drops (1 μL droplets) of ultrapure water, using a Laplace-Young fitting to extract the contact angle values of the drop on the surface.

*X-ray fluorescent (XRF).* The Cu concentration in the ELD layer was calculated from the XRF radiation intensity of each element recorded by a Spectro XEPOS analyzer on the basis of a stored set of internal standard calibration curves

*Anneal.* Coupon pieces with the following stack: Si/SiO2/APTS/ELD Cu received 10 min. anneal under forming gas atmosphere in a Heatpulse 610 rapid thermal process (RTP) annealer. A temperature range from 200°C to 450°C was spanned.

*Sheet resistance (Rs).* Cu silicide formation resulting from diffusion barrier failure could be detected visually by the darkened appearance of samples upon annealing [5]. Subsequently, Rs measurements were used to verify Cu diffusion into the substrate. Rs values after anneal were collected using an automatic four probe system from Four Dimensions Inc., model 280. A minimum of five measurements were collected for each coupon as a function of annealing temperature.

*Time-of-flight secondary ion mass spectroscopy (ToF-SIMS).* Chemical mapping using ToFSIMS has advantages in terms of spatial resolution (< 1 μm) and sensitivity (< 1 ppm) providing information on the chemistry of the surface at specific location of interest [10]. ToF-SIMS measurements were obtained using a commercial instrument Iontof-IV. Spectra were recorded in positive and negative ion mode from a Ga+ primary source operating at 15 KeV.

| Compound | Formula | Description | Concentration (M) |
|---|---|---|---|
| Dimethylamine borane (DMAB) | $(CH_3)_2NHBH_3$ | ≥ 97.0 %, Sigma-Aldrich | 0.1 |
| Boric acid | $H_3BO_3$ | ~ 99.0 %, Sigma-Aldrich | 0.05 |
| Copper (II) Chloride | $CuCl_2$ | ≥ 97.0 %, Sigma-Aldrich | 0.05 |
| Citric acid | $HOC(COOH)(CH_2COOH)_2$ | ≥ 99.5 %, Sigma-Aldrich | 0.15 |
| Tris(hydroxymethyl)aminomethane | $NH_2C(CH_2OH)_3$ | ≥ 99.8 %, Sigma-Aldrich | 0.5 |

## RESULTS AND DISCUSSION

*CA measurements.* Surface structure and morphology of the APTS layer deposited on the pre-cleaned silica surface were investigated as a function of silica type and pre-deposition treatment. No difference in the measured contact angle values after APTS deposition on a thermally grown $SiO_2$ (thermox) vs. a native $SiO_2$ (chemox) was observed. UV-ozone and pirañha cleaning ($H_2O_2$:$H_2SO_4$ 1:3) lead to CA respectively of 54.4° ± 3.0° and 55.2° ± 2.7°, which are in good agreement with published literature values of 60.0° ± 2.0° for fully amino-functionalized surfaces [11]. The presence of the APTS coating layer is revealed by an increase in the substrate RMS surface roughness from 0.148 nm on cleaned $SiO_2$ to 0.394 nm after functionalization (data not shown).

Nevertheless, UV-ozone treatment on the $SiO_2$ substrate before APTS deposition allowed for a CA value of 1.7° ± 1.5° lower than 15.0° ± 3.0° measured after pirañha cleaning.

*Optimization of the Cu ELD process parameters.* The purpose of this study is to explore and optimize the experimental conditions for a catalyst-free copper ELD process. Two main ELD mechanisms are evaluated.

In the first mechanism, copper particles are formed in the solution after a certain induction time that can be modified by adjusting experimental conditions, such as pH, temperature, ..... These negatively charged Cu/Cu oxide particles are supposed to electrostatically deposit on the posively charged APTS-functionalized surfaces, if the pH value is kept in the range 3-9 [7].

The second ELD mechanism consists in a "surface-nucleate growth', where Cu layer growth occurs in principle by in-situ condensation/reduction of the Cu-amino complexes.

None of the two mechanisms can be ruled out. Nevertheless, as observed by the naked eye, while the EL solution remained transparent showing the original blue color after mixing the raw materials, after a certain time the color changed to grass green and a brown precipitate finally appeared. The period from the beginning to the occurrence of precipitation within the solution, visible to the naked eye, is defined as the induction time ($\tau$). The induction time decreases with increasing pH of the solution, as evident from the SEM images in Fig. 1, where a Cu layer is observed after 30 min. only for pH values higher than 5. This phenomenon suggests that changing the pH can effectively control the oxidation–reduction reaction rate in solutions if the ionic concentrations are fixed. This is consistent with the electrochemistry of the oxidation reaction of the DMAB that is promoted at higher pH, so that the reduction of copper is accelerated.

The interfacial interaction between colloidal particles and flat surfaces has been studied in literature [12] and possible reactions are proposed, such as covalent coupling, and coordinative and electrostatic binding. Taking into account that copper/copper oxide particles and SAMs in the present study for pH lower than 9.5 are oppositely-charged [13], the electrostatic interactions can be likely regarded as one of the main factors, which is responsible for the formation of the copper micropatterns due to the selective adhesion.

Size and diffusion of the first Cu particles are controlled by the temperature and the Cu ion concentration in the EL solution. As evident from Fig. 2, for temperature below 50°C the formation of the Cu layer is inhibited, while for Cu ion concentration higher than 0.1 M, large Cu particles are obtained.

In addition, the reactions to form the Cu(II) and Cu(I) amino-complexes can not be neglected since both complexes have higher stability than the Cu(I)/citrate complex [7]. Nevertheless, on the basis of the reduction potentials of the copper ions and copper complexes that can be formed in our EL solution, the more stable complex ion $[Cu(NH_3)_4]^{2+}$ is the one that has the lowest ability to capture the electrons released from the oxidation of the DMAB. In other words, the preferential in-situ reduction of the copper ions fixed by the amino groups in the $NH_2$-SAM layer seems to be very unlikely.

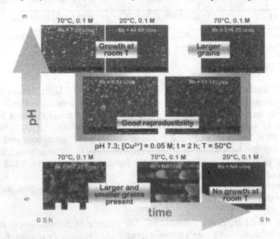

**Figure 1.** SEM images after the Cu ELD process as a function of pH and time.

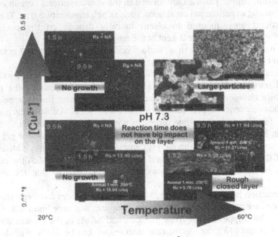

**Figure 2.** SEM images after the Cu ELD process as a function of $[Cu^{2+}]$ and temperature.

*ELD Cu nucleation and growth on clean SiO₂ and APTS-functionalized SiO₂ surfaces.* The SEM images shown in Fig. 3 compare the nucleation and growth of the ELD Cu layer on a) APTS-functionalized- and b) UV-ozone-cleaned- SiO$_2$ surfaces. The same silanization procedure has been initially applied in the two cases but, as shown in the schematics, in b) an UV-ozone step has been introduced followed by 30 min. Cu ELD. Cu nucleation on the APTS-functionalized surface shows much larger coverage (correspondent to a XRF Cu concentration of 3788 ppm) than that on the bare SiO$_2$ surface (correspondent to a XRF Cu concentration < 1 ppm) under the same optimized ELD experimental conditions, as reported in Table 1.

In addition, from the atomic concentrations of the Cu vs. Si background determined by XRF measurements for the different combination of the reaction parameters shown in Figs. 1 and 2, we can conclude that the optimized set of ELD experimental conditions leads to the maximum Cu/Si ratio on the APTS-functionalized surface (data not shown).

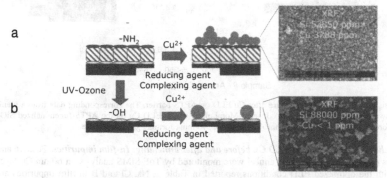

**Figure 3.** Schematics and SEM images after the optimized Cu ELD process a) with and b) without APTS layer at the interface between SiO$_2$ and Cu.

The easier Cu nucleation on the APTS - coated substrate than on the clean SiO$_2$ substrate can be explained on the basis of the already discussed mechanism. Cu shows a high mobility on the SiO$_2$ surface due to the weak interaction between Cu and O atoms, attributed to van der Waals forces acting at the Cu-SiO$_2$ interface [14].

On the contrary, the interaction between Cu and N atoms is strong enough to limit the migration of the Cu atoms resulting in a preferential positioning of Cu nuclei on the APTS surface. Thus, ELD Cu exhibits layer wetting. This wetting behaviour is accentuated upon annealing (10 min., 400°C, forming gas atmosphere), as confirmed by the SEM layer morphology shown in Figs. 3 a and b.

Ganesan *et al*. [15] reported that the interfacial debonding energy of the PVD Cu/NH$_2$ surface is ca. 5.5 J/m$^2$, which is higher than ca. 3.0 J/m$^2$ measured for the PVD Cu/SiO$_2$. This was considered an evidence for the immobilization of the Cu atoms in presence of the APTS functionalization. We measured the debonding energy of the ELD Cu/NH$_2$ surface that is ca. 2.2 J/m$^2$ before annealing, while it reaches ca. 4.5 J/m$^2$ after annealing [16]. This phenomenon confirms the enhancement of the ELD Cu layer wetting on APTS-functionalized surfaces upon temperature increase.

***Barrier properties assessment of the APTS layer.*** The APTS layer barrier properties and integrity were assessed by measuring the film resistivity ρ as a function of the anneal temperature in the range 200-450°C. An average ρ decrease of 22 μOhms cm is observed after anneal at temperatures in the range of 200 - 400°C. The observed ρ decrease upon anneal at 450°C is as low as 0.2 μOhms cm and the final ρ is higher with respect the final ρ values measured for anneal temperaures up to 400°C. In addition, the measured ρ values are very unstable for anneal temperatures above 450°C, as a confirmation that the Cu silicide formation is suppressed on NH$_2$-SAM up to anneal temperatures of 400-450°C [5]. Visual inspection of the samples after anneal reveals the presence of a black deposit only for anneal temperatures higher than 400-450°C, as an indication of Cu silicide formation due to barrier failure.

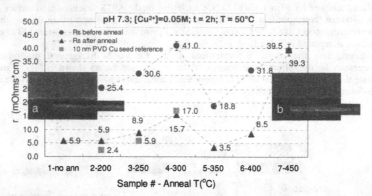

**Figure 4.** Resistivity vs. anneal temperature after Cu ELD on APTS barrier. The corresponding data from a blank sample is shown for comparison (squares). Inserts: Morphology of the ELD Cu on the APTS-functionalized $SiO_2$ surface a) before annealing, and b) upon annealing at 400°C, 10 min., forming gas atmosphere.

***ToF-SIMS characterization of the ELD Cu before and after annealing: In-film impurities.*** Positive and negative ions profiles before and after anneal were monitored by ToF-SIMS analysis on 60 nm Cu layer deposited with the optimized ELD conditions reported in Table 1. Na, Cl and B in-film impurities are detected. Impurities coming from the plating bath are one of the possible causes for the ten-fold higher $\rho$ observed for a 60 nm ELD Cu layer ($\rho \sim 30$ $\mu$Ohms cm) with respect to a PVD Cu film of equivalent thickness ($\rho \sim 3$ $\mu$Ohms cm). A decrease in the relative levels of Na and Cl impurities is observed upon anneal (400°C, 10 min., forming gas atmosphere).

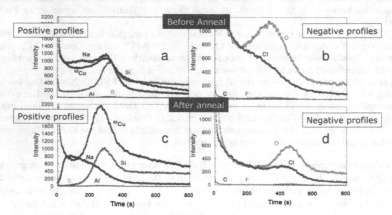

**Figure 5.** ToF-SIMS characterization of the ELD Cu layer: before anneal: a) positive profile, b) negative profile, and after anneal (400°C, 10 min., forming gas atmosphere): c) positive profile, d) negative profile).

## CONCLUSIONS

In this work, a convenient approach to the Cu ELD in the absence of any heterogeneous catalyst was studied. In particular, a wide range of experimental conditions such as pH, temperature, Cu ions concentration and time are investigated. The experimental results are used to validate proposed Cu ELD layer nucleation and growth mechanisms on the $SiO_2$ surface straightly after cleaning and after APTS functionalization. The APTS interface film is evaluated as a barrier layer to prevent copper diffusion into the underlying dielectrics by mean of a combination of multi-characterization techniques, such as CA, XRF, ToF-SIMS, SEM and $\rho$. The barrier properties of the APTS film after the optimized Cu ELD process are assessed up to ca. 400-450°C. The presence of copper particles in the EL solution is observed. Particle size, density and morphology can be controlled fine-tuning the ELD conditions and in particular pH and temperature. Two main nucleation and growth mechanisms were taken into account. In the first "particle attachment growth" mechanism, the electrostatic attraction between the negatively-charged copper colloids and the positive APTS surface (for pH below ca. 9.5) plays a role in first phase of the ELD Cu film formation. In the second "surface-nucleated growth" mechanism, the preferential in-situ condensation/reduction of copper ions fixed by the amino groups in the APTS layer is considered. Based on the relative reduction potentials of the Cu complexes formed in the studied "catalyst-less" ELD system and on experimental observations, the second mechanism seems less likely.

## ACKNOWLEDGMENTS

This work has been carried out within IMEC's Advanced Interconnect Industrial Affiliation Program. In particular, the authors gratefully thank G. Beyer, J. Coenen, M. De Raedt, A. Franquet, A. Gorrono, J. Steenbergen, Z. Tokei, and C. M. Whelan for their support.

## REFERENCES

[1] J. Chen, M. A. Reed, A. M. Rawlett, J. M. Tour, *Science*, **286**, 1550, (1999).
[2] J. J. Halls, C. A. Walsh, N. C. Greenham, E. A. Marseglia, R. H. Friend, S. C. Moratti, A. B. Holmes, *Nature*, **376**, 498, (1995).
[3] R. H. Friend, R. W. Gymer, A. B. Holmes, J. H. Burroughes, R. N. Marks, C. Taliani, D. D. Bradley D. A. Dos Santos, J. L. Bredas, M. Logdlund, W. R. Salaneck, *Nature*, **397**, 121, (1997).
[4] E. Glickman, N. Inberg, A. Fishelson, Y. Shacham-Diamand, *Microelec. Eng.*, **84**, 2466, (2007).
[5] A. Maestre Caro, G. Maes, G. Borghs, C. M. Whelan, *Microelec. Eng.*, **85**, 10, 2047, (2008).
[6] M. Yoshino, T. Masuda, T. Yokoshima, J. Sasano, Y. Shacham-Diamand, I. Maduda, T. Osaka, Y. Hagiwara, I. Sato, *J. Electrochem. Soc.*, **154**, 3, D122, (2007).
[7] P. Zhu, Y. Masuda, K. Koumoto, *J. Mater. Chem.*, **14**, 976, (2004).
[8] M. P. Hughey, D. J. Morris, R. F. Cook, P. Bozeman, P. B. Steven, L. K. Srinivas, L. N. Chakravarty, D. P. Harkens, L. C. Stearns, *Engineering Fracture Mechanics*, **71**, 2, 245, (2004).
[9] R. H. Dauskardt, M., Lane, Q. Ma, N. Krishna, *Engineering Fracture Mechanics*, **61**, 141, (1998).
[10] Y. K. Gong, K. Nakashima, *Langmuir*, **16**, 8546, (2000).
[11] A. Hozumi, Y. Yokogawa, T. Kameyama, H. Sugimura, K. Hayashi, H. Shirayama, O. Takai, *J. Vac. Sci. Technol. A*, **19**, 1812, (2001).
[12] C. S. Dulcey, J. H. Georger Jr., V. Krauthamer, D. A. Strenger, T. L. Fare, J. M. Calvert, *Science*, **252**, 551, (1991).
[13] J.-J. Shyue, Y. Tang, M. R. De Guire, *J. Mater. Chem.*, **15**, 323, (2005).
[14] Z.-Z. Liu, Q. Wang, X. Liu, J.-Q.- Bao, *Thin Solid Films*, **517**, 635, (2008).
[15] P. G. Ganesan, G. Cui, K. Vijayamohanan, M. Lane, G. Ramanath, *J. Vac. Sci. Technol. B*, **23**, 327, (2005).
[16] S. Armini, A. Maestre Caro, C. M. Whelan, *in preparation*, (2009).

Mater. Res. Soc. Symp. Proc. Vol. 1156 © 2009 Materials Research Society          1156-D04-09

# Rutherford Backscattering Spectrometry Analysis of Growth Rate and Activation Energy for Self-Formed Ti-Rich Interface Layers in Cu(Ti)/Low-k Samples

Kazuyuki Kohama[1], Kazuhiro Ito[1], Kenichi Mori[2], Kazuyoshi Maekawa[2], Yasuharu Shirai[1] and Masanori Murakami[1,3]

[1]Department of Materials Science and Engineering, Kyoto University, Kyoto 606-8501, Japan
[2] Process Tech. Develop. Div., Renesas Technology Corp., Itami-shi, Hyogo 664-0005, Japan
[3]The Ritsumeikan Trust, Nakagyo-ku, Kyoto 604-8520, Japan

## ABSTRACT

A new fabrication technique to prepare ultra-thin barrier layers for nano-scale Cu wires was proposed in our previous studies. Ti-rich layers formed at the Cu(Ti)/dielectric-layer interfaces consisted of crystalline TiC or TiSi and amorphous Ti oxides. The primary control factor for Ti-rich interface layer composition was the C concentration in the dielectric layers rather than the formation enthalpy of the Ti compounds. To investigate Ti-rich interface layer growth in Cu(Ti)/dielectric-layer samples annealed in ultra high vacuum, Rutherford Backscattering Spectrometry (RBS) was employed in the present study. Ti peaks were obtained only at the interface for all the samples. Molar amounts of Ti atoms segregated to the interface ($n$) were estimated from Ti peak areas. The $n$ value was defined by $n = Z \cdot \exp(-E/RT) \cdot t^m$, where $Z$ is a preexponential factor and $E$ the activation energy for the reaction. The $Z$, $E$, and $m$ values were estimated from plots of log $n$ vs log $t$ and log $n$ vs $1/T$. The $m$ values are similar in all the samples. The $E$ values for Ti atoms reacting with the dielectric layers containing carbon (except SiO$_2$) tended to decrease with decreasing C concentration (decreasing $k$), while reaction rate coefficients ($Z \cdot \exp(-E/RT)$) were insensitive to C concentration in the dielectric layers. These factors lead to conclusion that growth of the Ti-rich interface layers is controlled by chemical reactions of the Ti atoms with the dielectric layers represented by the $Z$ and $E$ values, rather than diffusion in the Ti-rich interface layers.

## INTRODUCTION

In our previous studies, a new fabrication technique to prepare ultra-thin barrier layers for nano-scale Cu wires was proposed. Supersaturated Cu(Ti) alloy films deposited on dielectric layers such as SiO$_2$ and SiO$_x$C$_y$ with low dielectric constants (Low-k) were annealed at elevated temperatures, and thin Ti-rich layers were found to be formed at the interfaces [1,2]. The Ti-rich layers formed at the interfaces were found to consist of crystalline TiC or TiSi in addition to amorphous Ti oxides. The primary factor to control composition of the self-formed Ti-rich interface layers was the C concentration in the dielectric layers rather than the formation enthalpy of the Ti compounds (TiC and TiSi). Crystalline TiC was formed on the dielectric layers with a C concentration higher than 17 at.% [2].

Transmission electron microscopy (TEM) observation and electrical resistivity measurements indicated that growth of the Ti-rich interface layers consisting of TiC (Cu(1 at.%Ti)/Low-k samples) was faster than those consisting of TiSi (Cu(1at.%Ti)/SiO$_2$ samples) after annealing at 400°C [3]. For systematic investigation of dependence of initial Ti contents in the Cu(Ti) alloy films and C concentration in the dielectric layers on the Ti-rich interface layer growth, Rutherford Backscattering Spectrometry (RBS) technique was employed instead of TEM in the present studies.

## EXPERIMENTAL PROCEDURES

The Cu(1, 5, and 10 at.%Ti) alloy films were deposited on $SiO_xC_y$ (Low-k1 and Low-k4), $SiO_2$, SiCO, and SiCN dielectric layers by a radio frequency magnetron sputtering technique. The dielectric constants ($k$) and compositions of these layers are shown in Tabel 1. The base pressure prior to deposition was approximately $1 \times 10^{-6}$ Pa, and the sputtering power and working pressure were kept at 300 W and about 1 Pa, respectively. The typical thickness of the Cu(Ti) alloy films were controlled to approximately 400 nm. The samples were annealed in ultra high vacuum (UHV) at 400°C~650°C for 2 h~72 h. The film microstructures were analyzed by X-ray diffraction (XRD) and TEM. The Ti segregation to the interface was investigated by RBS. For the RBS measurements, $^4He^{2+}$ ion beams with energy of 2 MeV were impinged perpendicularly onto the film surfaces.

**Table 1**. Dielectric constants $k$ of the dielectric layers and their compositions (at.%).

|        | $k$ | C (at.%) | O (at.%) | Si (at.%) | N (at.%) |
|--------|-----|----------|----------|-----------|----------|
| SiCN   | 4.8 | 21.4     | 0.5      | 25.0      | 12.8     |
| SiCO   | 4.5 | 20.8     | 16.6     | 24.6      | -        |
| Low-k1 | 3.0 | 17.0     | 24.9     | 18.8      | -        |
| Low-k4 | 2.6 | ~14      | ~29      | ~18       | -        |
| SiO₂   | 3.9 | -        | 66.7     | 33.3      | -        |

## THEORY

A peak area ($A$) of an element in RBS profiles is the product of the incident beam dose ($Q$), the number of the element atom in a unit of area ($N$), the scattering cross section of the element atom ($\sigma$), and the solid angle of the detector ($\Omega$) [4]:

$$A = QN\sigma\Omega .\qquad(1)$$

The molar amount of Ti atoms segregated to the interface ($n$) was given by dividing their number ($N$) from eq. (1) by Avogadro's number $N_A$:

$$n = N/N_A = A/N_A Q\sigma\Omega .\qquad(2)$$

In the present study, each Ti peak was fitted by a Gaussian curve and an error function, and the $A$ value was determined from the area under the fitted curve. The values of Q, $\Omega$, and $\sigma$ of Ti atoms were $3.12 \times 10^{13}$ (10 μC), $3 \times 10^{-3}$ (steradian) and $6.28 \times 10^{-29}$ $m^2$ (0.628 barn), respectively.

Figure 1(a) shows RBS profiles of a Cu(1at.%Ti)/Low-k1 sample before and after annealing at 400°C for 2h in UHV. A Ti peak was observed only at the interface. Similar profiles were obtained in all the Cu(Ti)/dielectric-layer samples after annealing in UHV. Figure 1(b) shows a portion of the RBS profile of Cu(1at.%Ti)/Low-k1 sample after annealing at 400°C for 2h in UHV (Fig. 1(a)), and a refinement plot (solid line) was placed upon the observed data. The observed RBS profile was fitted by the sum of three components: Ti segregation at the interface, Ti atoms in the alloy film, and a Cu edge above the Ti segregation at the interface. A best fit of the calculated pattern to the observed pattern was obtained on the basis of various indicators. One of the important indicators is the weighted pattern $R$ index, $R_{wp}$, defined as

$$R_{wp} = \{\sum w_i(y_i(obs) - y_i(calc))^2 / \sum (w_i(y_i(obs))^2\}^{1/2}, \qquad (3)$$

where $w_i = 1/y_i(obs)$. The numerator of the expression inside the braces is the sum of squared residuals and thus has expected value $o-p$, where $o$ and $p$ are the number of observations and the number of parameters, respectively. The expected $R$ index, $R_e$, is accordingly defined as

$$R_e = \{o-p/\sum (w_i(y_i(obs))^2\}^{1/2}. \qquad (4)$$

The ratio $R_{wp}/R_e$ (=$S$) is a measure of how well the fitted model accounts for the data [5]. The $S$ value of 1.24 is usually considered to be quite satisfactory ($1.0<S<1.3$). The RBS profiles of Cu(Ti)/dielectric-layer samples after annealing in UHV were similarly fitted, and the $S$ values obtained were in the range of 1.0 to 1.3 in all the samples.

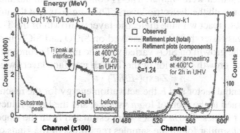

**Figure 1.** (a) RBS spectrum profiles of the Cu(1at.%Ti)/Low-k1 samples before and after annealing at 400°C for 2h in UHV. (b) A portion of the RBS profile (a) and refinement plot (solid line) placed upon the observed data.

## RESULTS AND DISCUSSION

### Comparison of Growth Analyses Using RBS and TEM

Growth of the Ti-rich interface layers on the dielectric layers after the annealing is exhibited as $n$ values estimated from the $A$ values originating in the Ti peaks. Plots of $n$ vs annealing time ($t$) for Cu(1 at.%Ti)/dielectric-layer samples are shown in Fig. 2(a). For comparison, the Ti-rich interface layer thicknesses obtained by TEM are shown in Fig. 2(b) [3].

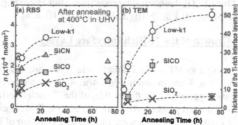

**Figure 2.** Plots of $n$ vs annealing time ($t$) for Cu(1 at.%Ti)/dielectric-layer samples after annealing at 400°C in UHV observed by (a)RBS and (b)TEM.

The growth of the Ti-rich interface layers on Low-k1 was the fastest, followed by those on SiCO or SiCN, while that on $SiO_2$ was the lowest in both the figures. These factors indicate that an RBS technique is an appropriate method for the growth analysis of the Ti-rich interface layers.

## Growth Behavior of the Ti-Rich Interface Layers

Figures 3(a)-3(c) show the growth behaviors of the Ti-rich interface layers after annealing at 400°C for the inital Ti concentrations of 1, 5, and 10 at.%, respectively. The $n$ values increased with increasing annealing time and gradually arrived at saturating values in all the samples. The $n$ values for the Cu(1 at.%Ti) samples were lower than those of the Cu(5at.%Ti) and Cu(10at.%Ti) samples, which were similar. The Ti concentration in the alloy films continued to decrease during reaction with the dielectric layers. This suggests that the initial Ti concentration of 1 at.% is insufficient to react with the dielectric layers. Growth of the Ti-rich interface layers observed on Low-k1 was faster than those observed on SiCO and SiCN in all the initial Ti concentrations, while the growth on $SiO_2$ varied with the initial Ti concentration.

On the other hand, growth behavior of the Ti-rich interface layers was defined by

$$n = Z \cdot \exp(-E/RT) \cdot t^{m},$$ (5)

where $Z$ is a preexponential factor and $E$ the activation energy for the reaction. The plots of $n$ vs $t$ (Figs. 3) were transformed to the plots of log $n$ vs log $t$, and the log $n$ values were found to be proportional to the log $t$ values. The $m$ values were estimated from the slopes of the log $n$ - log $t$ lines. They were almost similar for all the samples regardless of the kinds of dielectric layers at each initial Ti concentration in the alloy films, suggesting similar growth mechanism.

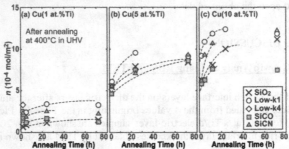

**Figure 3**. Plots of $n$ vs $t$ for Cu(Ti)/dielectric-layer samples after annealing at 400°C in UHV. The initial Ti concentrations are (a) 1 at.%, (b) 5 at.%, and (c) 10 at.%.

## Activation Energies of Growth of the Ti-Rich Interface Layers

Arrhenius plots of log $n$ vs $1/T$ in Cu(Ti)/dielectric-layer samples for the inital Ti concentrations of 1, 5 and 10 at.% are shown in Figs. 4(a)-4(c), respectively. They showed a linear relationship, and this suggests that the formation of the Ti-rich interface layers was controlled by a thermally-activated process. Activation energies, $E$, were estimated from the slopes of the log $n$ vs $1/T$ lines. The $E$ values were plotted as a function of C concentration of the dielectric layers, as shown in Fig. 5(a). The $E$ values in the initial Ti concentration of 1at.% were lower than those at 5at.%Ti and 10at.%Ti, which were similar. The initial Ti concentration of 1at.% is insufficient to induce reaction of the Ti atoms with the dielectric layers, as mentioned earlier, and thus the $E$ values in the initial Ti concentrations of 5at.% and 10at.% are concluded to be an appropriate activation energies of the formation of the Ti-rich interface layers.

The $E$ values for the samples consisting of dielectric layers containing carbon (except $SiO_2$) tended to decrease with decreasing C concentration (decreasing $k$), and those without carbon ($SiO_2$) were much higher than others. This indicates that the reaction of the Ti atoms in the alloy films with the Low-k layers is the easiest thermally-activated process. Also, composition of the dielectric layers is suggested to play an important role in the reaction of the Ti atoms with dielectric layers, and the carbon may be a key element to control the reaction. This is similar to the formation rule of Ti compounds (TiC or TiSi) in the Ti-rich interface layers [2]. The preexponential factors, $Z$, estimated from intercepts of the slopes in the log $n$-log $t$ and log $n$ - $1/T$ lines were the same, and similarly varied with C concentration in the dielectric layers. The $Z$ value shows the frequency with which the Ti atoms meet elemental reactants in the dielectric layers. The frequency was low in the samples with Low-k, and was high in the samples with $SiO_2$. In consequence, reaction rate coefficients ($Z \cdot exp(-E/RT)$) were insensitive to C concentration in the dielectric layers at the initial Ti concentrations of 5 at.% and 10 at.%, while those increased with decreasing C concentration at the initial Ti concentrations of 1 at.%.

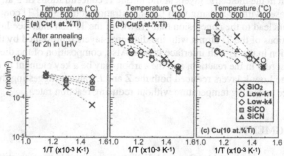

**Figure 4**. Arrhenius plots of log $n$ vs $1/T$ for Cu(Ti)/dielectric-layer samples after annealing for 2 h in UHV. The initial Ti concentrations are (a) 1 at.%, (b) 5 at.%, and (c) 10 at.%.

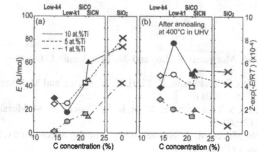

**Figure 5**. (a) Activation energies of $E$ and (b) values of $Z \cdot exp(-E/RT)$ of the Cu(Ti)/dielectric-layer samples after annealing in UHV, as a function of C concentration of the dielectric layers.

Those factors lead to the conclusion that growth of the Ti-rich interface layers is controlled by chemical reactions of the Ti atoms with the dielectric layers represented by the $Z$ and $E$ values, rather than diffusion in the Ti-rich interface layers. Also, composition of the dielectric layers plays an important role in the reaction, and carbon may be a key element to control the reaction.

## CONCLUSIONS

For systematic investigation of growth of the Ti-rich interface layers in the annealed Cu(Ti)/dielectric-layer samples, the RBS technique was employed. In all the samples after annealing in UHV, Ti peaks were obtained only at the interface in the RBS profiles. The $n$ values estimated from the $A$ values increased with increasing annealing time and temperature. The log $n$ values were found to be proportional to the log $t$ values. The slopes were almost similar for all the samples, suggesting similar growth mechanism. Activation energy ($E$) for the reaction of the Ti atoms with the dielectric layers containing carbon (except $SiO_2$) tended to decrease with decreasing C concentration (decreasing $k$), and those without carbon ($SiO_2$) were much higher than others. The preexponential factor ($Z$) values similarly varied with C concentration in the dielectric layers. In consequence, coefficients of the reaction rate ($Z \cdot exp(-E/RT)$) were insensitive to C concentration in the dielectric layers at the initial Ti concentrations of 5 at.% and 10 at.%, while those increased with decreasing with C concentration at the initial Ti concentration of 1 at.%. Those factors lead to the conclusion that growth of the Ti-rich interface layers is controlled by chemical reactions of the Ti atoms with the dielectric layers represented by the $Z$ and $E$ values, rather than diffusion in the Ti-rich interface layers. Also, composition of the dielectric layers plays an important role in the reaction, and the carbon may be a key element to control the reaction. Using the Low-k layers reduced both the $Z$ and $E$ values, suggesting that there is some potential to reduce annealing temperature without reducing reaction rate.

## ACKNOWLEDGMENTS

This work was supported by Research Fellowships of the Japan Society for the Promotion of Science for Young Scientists (Kohama) and Grants-in-Aid for the Global COE Program from the Ministry of Education, Culture, Sports, Science and Technology of Japan.

## REFERENCES

1. S. Tsukimoto, M. Moriyama K. Ito, and M. Murakami, *J. Electron. Mat.*, **34**, 592-599 (2005).
2. K. Kohama, K. Ito, S. Tsukimoto, K. Mori, K. Maekawa, and M. Murakami, *J. Electron. Mat.*, **37**, 1148-1157 (2008).
3. K. Kohama, K.Ito, S. Tsukimoto, K. Mori, K. Maekawa, and M. Murakami, *Mat. Trans.*, **49**, 1987-1993 (2008).
4. W.-K. Chu, J.W. Mayer, M.-A. Nicolet, Backscattering Spectrometry, San Diego : Academic Press, 1978, pp. 91-92.
5. R.A. Young, The Rietveld Method, New York: Oxford University Press, 1993, pp. 22.

Mater. Res. Soc. Symp. Proc. Vol. 1156 © 2009 Materials Research Society     1156-D04-10

# Adhesion and Cu Diffusion Barrier Properties of a MnO$_X$ Barrier Layer Formed With Thermal MOCVD

K. Neishi[1], V. K. Dixit[1], S. Aki[1], J. Koike[1]
K. Matsumoto[2], H. Sato[2], H. Itoh[2], S. Hosaka[2]
[1]Department of Material Science, Tohoku University, Sendai 980-8579, Japan
[2]Technology Development Center, Tokyo Electron Ltd., Nirasaki, 407-0192, Japan

## Abstract

A thin and uniform manganese oxide layer was formed below 400 °C by thermal chemical vapor deposition (CVD) on a tetraethylorthosilicate (TEOS) oxide substrate. No Cu delamination were observed both in the as-deposited and in the annealed PVD-Cu/CVD-MnO$_x$/TEOS samples deposited below 300 °C, meanwhile the Cu films were delaminated from the CVD-MnO$_x$/TEOS substrates deposited at 400 °C. From the results of XPS, Raman and SIMS analysis, a major reason for degradation of the adhesion properties is considered to be amount of carbon inclusion in the CVD Mn oxide layer.

## Introduction

With the increase in packing density of ultra-large scale integrated circuits, geometrical dimensions continue to decrease. As the width of Cu line becomes narrow, the thickness of a barrier layer should be reduced to prevent the increase of the effective resistivity of the interconnect lines [1]. The barrier layer formation with a physical vapor deposition (PVD) technique has become difficult as the technology node is reduced to 45 nm because of poor step coverage.

Chemical vapor deposition (CVD) has been known for its good step coverage, and is a candidate technique for the formation of a barrier layer, a Cu seed layer and entire Cu line. In our previous research, we investigated the formation behavior of manganese oxide layer with thermal CVD and the diffusion barrier properties of the oxide [2,3]. The thin and uniform amorphous MnO$_x$ layer could be formed on a blanket TEOS substrate as well as in a contact-hole structure and a dual damascene structure. Transmission electron microscopy indicated a good diffusion barrier property after annealing at 400 °C for 100 h. In addition to the reported properties, a good adhesion strength is necessary between a Cu line and a dielectric layer not only to ensure good SM and EM resistance but also to prevent film delamination under mechanical or thermal stress conditions during fabrication process such as chemical mechanical polishing or high temperature annealing. To date, no information is available with regard to the adhesion property of CVD-MnO$_x$ barrier layer.

In this work, we investigated adhesion and diffusion barrier properties in further detail using X-ray photoelectron spectroscopy (XPS), secondary ion mass spectroscopy (SIMS) and Raman spectroscopy. The dependence of the adhesion property on deposition temperature was correlated with the chemical composition and valence state of Mn and with the amount of carbon contained in the MnO$_x$ layer.

## Experimental procedures

Substrates were p-type Si wafers having plasma TEOS oxide of 100nm in thickness. Bisethylcylopentadienyl manganese (EtCp)$_2$Mn was used as a metal organic manganese precursor. The Mn precursor was vaporized at 80 °C and introduced into the deposition chamber with $H_2$ carrier gas to deposit the $MnO_x$ barrier layer by thermal-CVD process at 100-400 °C. Subsequently, the sample was transferred to a load-lock chamber where a DC sputtering system was employed to deposit a Cu overlayer at room temperature to a thickness of 100 nm. Two types of the samples were prepared in this study. One is a sample without a Cu overlayer to analyze the chemical bonding state of the $MnO_x$ layer. The other is a sample with a Cu overlayer to investigate adhesion and diffusion barrier properties between the $MnO_x$ layer and the Cu overlayer. To investigate the diffusion barrier property, the sample was annealed at 400 °C for 100 hours in a vacuum of better than $1.0 \times 10^{-5}$ Pa.

Transmission electron microscopy (TEM) attached with X-ray energy dispersive spectroscopy (EDS) was used to investigate the microstructure, thickness of the $MnO_x$ layer and chemical composition of each layer. XPS was used to analyze the chemical bonding state of the CVD-Mn layer. SIMS was used to analyze the depth profile of each elements. Raman spectroscopy was used to analyze the surface state.

### Results and Discussion

Table 1 shows the result of Scotch tape test for adhesion property in the as-deposited and the annealed samples of $Cu/MnO_x/SiO_2/Si$. Temperatures at the top raw of the table indicate the deposition temperature of the $MnO_x$ layer. Annealing was performed at 400 °C for 100 hours in vacuum. No copper films are delaminated in the as-deposited and the annealed samples when the $MnO_x$ layer was deposited below 300 °C. On the other hand, the Cu films are delaminated in the sample deposited at 400 °C.

**Table 1.** The result of Scotch tape test to the as-deposited and annealed $Cu/CVD-MnO_x/SiO_2/Si$ samples.

| Deposition temp. | 100 °C | 200 °C | 300 °C | 400 °C |
|---|---|---|---|---|
| As-dep. | No peel | No peel | No peel | peel |
| Annealing at 400 °C for 100h | No peel | No peel | No peel | peel |

Figure 1 shows the Raman spectra normalized to the peak intensity of crystalline Si at 515 cm$^{-1}$ in the as-deposited $MnO_x/SiO_2/Si$ samples and the $SiO_2/Si$ reference sample. No obvious peaks of $MnO_x$ can be observed in the $MnO_x/SiO_2/Si$ samples compared to the $SiO_2/Si$ reference sample. The Raman spectra of the sample deposited at 400 °C shows broad peaks around 1000, 1400 and 1600 cm$^{-1}$. The peak around 1000 cm$^{-1}$ corresponds to the crystalline Si. Other peaks are in agreement with amorphous carbon peaks of D (1357 cm$^{-1}$), G (1582 cm$^{-1}$) and D' (1620 cm$^{-1}$) bands [4,5]. In contrast, the carbon related peaks are not noticeable in the $MnO_x$ layer deposited below 300 °C. This result implies that the amorphous carbon is introduced from the fragment of the Mn precursor to the $MnO_x$ layer during deposition at 400 °C and these carbons are contained in the $MnO_x$ layer or adsorbed at the surface of the $MnO_x$ layer.

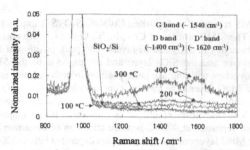

**Figure 1.** The Raman spectra of the as-deposited CVD-MnO$_x$/SiO$_2$/Si samples and the SiO$_2$/Si sample.

Figure 2 shows the XPS spectra of C 1s of the as-deposited MnO$_x$/SiO$_2$/Si samples deposited at 300 °C and 400 °C, respectively. Below 300 °C, the main peak of C-C and the peaks of C-O/C=O and carbidic carbon exist as shown in the XPS spectra of 300 °C sample. On the other hand, the main peak of the carbidic carbon and the peaks of C-C exist in the sample deposited at 400 °C. It is clear that the presence of the carbon contributes the poor adhesion. However, quantitative determination of the C-C bonds is difficult due to carbon contamination of the film surface from air exposure to the sample. Here, the peak integral intensity ratio of I$_{carbidic}$/ I $_{carbon}$ $_{total}$ is compared. Peak analysis indicates that the intensity ratio of the sample deposited at 400 °C is higher than that deposited below 300 °C. These XPS results show that the amount of the carbidic carbon in the MnO$_x$ layer deposited at 400 °C is higher than that deposited below 300 °C. Thus, it is considered that a large number of the Mn-C bonds (Mn-EtCp state) are formed at 400 °C.

Additionally, the valence state of the MnO$_x$ layer deposited below 400 °C is found to be Mn$^{2+}$ and the oxide stoichiometry is MnO based on the peak-position analysis of Mn 2p$_{3/2}$, the peak spacing between the Mn 2p$_{3/2}$ satellite and the Mn 2p$_{3/2}$, and the splitting of Mn 3s [6]. Hereafter, the MnO$_x$ formed below 400 °C will be denoted MnO. It is noted that this valence state is in agreement with that of the self-formed Mn oxide barrier layer (MnO+MnSiO$_3$) on the porous-SiOC with the PVD Cu-Mn alloy film [7].

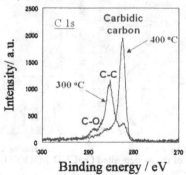

**Figure 2.** The XPS spectra of C 1s in the as-deposited CVD-MnO$_x$/SiO$_2$/Si samples deposited at 300 and 400 °C.

Figure 3 shows SIMS results of as-deposited Cu/MnO/SiO$_2$/Si samples formed at 200 °C and 400 °C, respectively. The SIMS profiles of Cu, Mn, Si, O and C were obtained by etching from the Cu surface side through Cu, SiO$_2$ and Si layers. The peak position of the Mn profile exists between the Cu overlayer and the SiO$_2$ layer and corresponds to the MnO layer. The peak position of C profile overlaps with the Mn peak position in both samples and corresponds to the existence of C in the MnO layer. The carbon concentration in the MnO layer of the sample deposited at 400 °C is about 5 times higher than that of the sample deposited at 200 °C.

From the Raman, XPS and SIMS results, it is considered that a major reason for the poor adhesion in the sample deposited at 400 °C is more inclusion of the amorphous-like carbon and carbidic carbon in the MnO layer compared to samples deposited below 300 °C.

The pyrolysis of the (EtCp)$_2$Mn precursor is reported to occur at temperature between 400 and 500 °C [2,8]. Below the decomposition temperature of the cyclopentadienyl precursors of Cp$_2$Ni and Cp$_2$Fe, it is reported that the precursor is adsorbed on the metal substrate and undergo decomposition to cyclopentadiene under electron or photon irradiation [9-11]. The Mn precursor gas is adsorbed on the substrate and interacts with the Si-OH/Si-O bonds at the surface of SiO$_2$ at the un-decomposition temperature of 300 °C. As a result, it is regarded that the MnO layer and (EtCp)H gas are formed to a larger extent and the Mn-(EtCp) is formed to a lesser extent. Meanwhile, above the decomposition temperature of 400 °C, the part of the Mn precursor thermally decomposes to Mn atoms and reactive (EtCp) ligands. The decomposed species react with the substrate and form the moderate amount of the carbidic carbon of (EtCp)-Mn compared to the case below 300 °C. Moreover, it is reported that the decomposition of the absorbed Cp ligands on Ag substrate resulted in the formation of carbon atoms [12]. Therefore, the poor adhesion of the 400 °C sample is due to the presence of amorphous-like carbon as well as the carbidic carbon in the form of EtCp-Mn. Both carbon types are expected to have weak bonding with Cu.

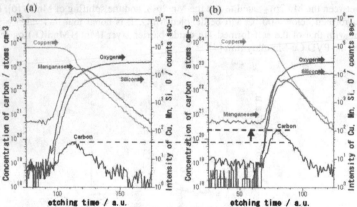

**Figure 3.** The SIMS results of the as-deposited Cu/CVD-MnO/SiO$_2$/Si samples deposited at (a)200 and (b)400 °C.

Figure 4 shows SIMS profiles of the annealed $Cu/MnO/SiO_2/Si$ sample deposited at 100 °C. The SIMS profiles of Cu, Mn, Si, O and C obtained by etching from the backside through the Si layer, the $SiO_2$ layer and the Cu overlayer. In our previous work, TEM observation indicated that the MnO layer remained to be a thin amorphous structure, and TEM-EDS spectra of TEOS layer indicated no Cu diffusion [2]. The Cu SIMS profile across the MnO layer to the $SiO_2$ layer shows sharp decline at the Cu/MnO/TEOS region. This result confirms no Cu diffusion into the $SiO_2$ layer after longtime annealing and is in agreement with the previous results.

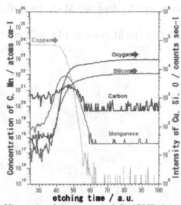

**Figure 4.** The back-side SIMS result of the annealed $Cu/CVD-MnO/SiO_2/Si$ sample deposited at 100 °C. Heat treatment is performed at 400 °C for 100 hours.

## Conclusions

The thin and amorphous MnO layer with the thermal-CVD was formed on the TEOS-$SiO_2$ below 400 °C. Good adhesion strength between the Cu and the MnO layers can be obtained when the MnO was deposited below 300 °C. It is considered that the degradation of the adhesion strength is due to the larger extent of carbon contamination at the higher deposition temperature. The thin and amorphous MnO layer has an excellent diffusion barrier property. The CVD-MnO process is promising technique to form a conformal and reliable barrier layer for an advanced technology node.

## Acknowledgements

We thank N. Akao of Tohoku University for his technical supports and suggestions for XPS analysis.

## References

1. International Technology Roadmap for Semiconductors, 2006.
2. K.Neishi, S. Aki, K. Matsumoto, H. Sato, H. Ito, S. Hosaka and J. Koike, Appl. Phys. Lett., **93**, 032106(2008).
3. K. Matumoto, K. Neishi, H. Ito, H. Sato, S. Hosaka and J. Koike, Appl. Phys. Express, **2**, 036503(2009).

4.  F. Tuinstra and J. L. Koenig, J. Composite mater., **4**, 492(1970).
5.  T. Jawhari, A. Roid and J. Casado, Carbon, **33**,1561(1995).
6.  V. K. Dixit, K. Neishi, J. Koike, Proceedings of the MRS spring, pp D4-11(2009).
7.  J. M. Ablett, J. C. Woicik, Zs. Tőkei, S. List and E. Dimasi, Appl. Phys. Lett., **94**, 042112(2009).
8.  S. Wen-bin, K. Durose, A. W. Brinkman and B. K. Tannor, Material Chemistry and Physics, **47**, p.75-77(1997).
9.  D. Welipitiya, A. Green, J. P. Woods and P. A. Dowben, J. Appl. Phys., **79**, 8730(1996).
10. D. L. Pugmire, C. M. Woodbridge and M. A. Langell, Sur. Sci., **411**, L844(1998).
11. D. L. Pugmire, C.M. Woodbridge, S. Root and M. A. Langell, J. Vac. Sci. Technol. A, **17**, 1581(1999).
12. D. L. Pugmire, C.M. Woodbridge, N. M. Boag and M. A. Langell, Sur. Sci., **472**, 155(2001).

Mater. Res. Soc. Symp. Proc. Vol. 1156 © 2009 Materials Research Society          1156-D04-11

## Electronic Transport Properties of Cu/MnO$_X$/SiO$_2$/p-Si MOS Devices

V. K. Dixit, K. Neishi and J. Koike

Department of Material Science, Tohoku University, Sendai 980-8579, Japan

## ABSTRACT

An ultrathin barrier layer of MnO$_x$ was grown using metal organic chemical vapor deposition (MOCVD) at an interface between Cu and SiO$_2$ dielectric. The electronic transport properties of Cu/MnO$_x$/SiO$_2$/p-Si metal oxide semiconductor (MOS) devices showed leakage current density within the range of $10^{-8}$-$10^{-7}$A/cm$^2$ up to an electric field of 4MV/cm. The current density remained within the same range after bias temperature aging test at 3MV/cm for $6\times10^3$s at 550K. The capacitance-voltage curves of the MOS device having the MnO$_x$ layer grown at 473K do not show significant shift of flat band voltage after thermal annealing at 673K for $3.6\times10^3$s as well as after bias temperature aging test at 1MV/cm, 550K for $2.4\times10^3$ s. These results indicate that the ultrathin layer of MnO$_x$ is stable under the above conditions and prevents sufficiently Cu ion diffusion into the SiO$_2$ dielectric.

## INTRODUCTION

As the technology node of silicon semiconductors is advanced to a nanometer range, resistance capacitance (RC) delay becomes a serious issue [1-2]. In order to maintain line resistance as low as possible, Cu has been used as the interconnect metal. Since interdiffusion occurs easily between Cu and SiO$_2$ dielectric insulating layer, a diffusion barrier layer is necessary at the interface. However, the barrier layer is not only a high resistivity material but also takes up the portion of the Cu line volume, leading to the increase of effective line resistivity. Thus, the barrier layer should be as thin as possible. Conventional material for the diffusion barrier is a bilayer structure of Ta/TaN formed by sputter deposition. Because of the straight trajectory of energetic sputtered atoms, the barrier layer becomes thinner on the side-wall surface of line trenches and connecting via holes than on other surfaces facing directly towards the sputter cathodes. This thickness non-uniformity of the conventional barrier has been a bottle neck for further thickness reduction.

Recently, we developed chemical vapor deposition (CVD) process to form a conformal layer of a thin Mn oxide having a thickness of ~2 nm [3]. The diffusion barrier property was confirmed by an x-ray energy dispersive spectrometer (EDS) attached to a transmission electron microscope. However, the EDS analysis is not capable of detecting the small amount of secondary elements of less than 0.1 at.%. Since electric properties of semiconductors are sensitive to the ppm level of impurities, the detection accuracy of EDS is not sufficient to ensure the diffusion barrier property for interconnect application. In the present work, we investigated the barrier property of the MnO$_x$ ultrathin layer at Cu/SiO$_2$ interface through current-voltage (I-V) and capacitance-voltage (C-V) measurement. The charge migration into Cu/MnO/SiO$_2$/p-Si/Al MOS devices and its effects on the electrical behavior are also addressed through thermal annealing and bias temperature aging (BTA) treatment at different temperatures for different periods of time.

## EXPERIMENTAL PROCEDURES

The Substrates were thermally grown $SiO_2$ or plasma TEOS-$SiO_2$ on p-type Si wafers. The p-TEOS/p-Si substrates were provided by TOSHIBA. A very thin barrier layer of $MnO_x$ was grown on these substrates by MOCVD at the growth temperature ($T_G$) of 373 to 773 K. Bisethylcyclopentadienyl manganese ((Et Cp)$_2$Mn) precursor was used as a metal organic precursor [3]. After the growth of $MnO_x$, a Cu overlayer was deposited to 150 nm thick, using a DC sputtering system located in a load-lock chamber. This enabled Cu overlayer deposition without exposing the $MnO_x$ barrier layer to the air. Secondary ion mass spectroscopy (SIMS) was performed to investigate the depth profile of elements in the PVD-Cu/CVD-MnO/$SiO_2$/Si samples. The Cs ion was used as a primary ion with an accelerating voltage of 3.0 keV. The secondary ion signal was obtained from the area of 60 x 60μm$^2$. The blanket Cu layer samples were patterned to an electrode pad array of 50 x 50μm$^2$, 60 x 120μm$^2$, or 100 x 100μm$^2$ dimensions with standard photolithography process and subsequent wet or dry etching. Wet etching of Cu was carried out in a $H_2O_2$ based solution provided by Mitsubishi Gas Chemicals. Dry etching of Cu was carried out by fast atom beam (FAB) etching using $SF_6$ gas at a flow rate of 5.6 sccm, the discharge voltage of 1.97 kV and the discharge current of 15mA for optimum condition. The back-side Ohmic contact of the p-Si wafer was formed with Al-1at%Si of 150nm thick. The samples were annealed at different temperatures of 573, 673 and 773 K for 3.6x10$^3$ s in an infrared lamp furnace under the vacuum of better than 1.0x10$^{-5}$ Pa. The frequency dependent C-V data of Cu/MnO/$SiO_2$/p-Si/Al MOS devices were measured using a microprobe system connected to the Agilent 4155C parametric analyzer and the Solartron 1260 impedance/gain-phase analyzer. Bias temperature aging (BTA) studies were carried out in a microprobe chamber at the vacuum of 1.0x10$^{-3}$ Pa with controlled set temperature having an accuracy of ±1K. The uniform temperature on the sample was achieved by using a Pt coated AlN heating plate.

## RESULTS AND DISCUSION

Figures 1a and b show the SIMS results of Cu/$MnO_x$/$SiO_2$/p-Si for different $T_G$ of $MnO_x$. The Mn concentration at the interface of Cu/$SiO_2$ increases with growth temperature. Though not shown here, a good adhesion between Cu and MnOx was observed by the tape test of the samples grown below 637 K. The poor adhesion above 673 K was due to the increased pyrolysis of (EtCp)$_2$Mn.

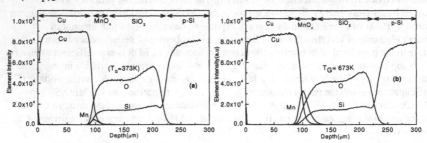

**Fig. 1.** SIMS results of as-deposited Cu/$MnO_x$/$SiO_2$/p-Si at a) 373K, b) 673K

106

Raman spectra also confirmed the increase of carbon content with increasing $T_G$ [4]. This is either carbon amount in the layer is more due to its thickness or carbon incorporation is high with increasing $T_G$. The valence state of Mn in $MnO_x/SiO_2$ was +2 as investigated by exchange splitting of various core levels of Mn using x-ray photoelectron spectroscopy (XPS) [4]. The valence-band offsets between the ultrathin MnO layer and the $SiO_2$ dielectric was $2.0\pm0.1$ eV as determined by XPS[4]. However its complete band structure is under investigation in combination with inverse photoelectron spectroscopy analysis. The thickness of the MnO layer depends on the growth temperature and varies from 2 to 8 nm from 373 to 673 K [3]. In this work, we used the MnO barrier layer grown at 473K on $SiO_2$/p-Si for MOS device fabrication.

Figure 2(a) shows the current density-electric field (*J-E*) curves of the $Cu/MnO/SiO_2$/p-Si MOS devices subjected to different treatment conditions. The current density is significantly large for the MOS devices that were processed using dry (FAB) etching. This is possibly due to ion damage that was caused by high energy $SF_6$ ions. Hence this concluded the dry etching of Cu on $MnO/SiO_2$ is not appropriate. In contrast, the MOS samples processed with wet etching indicate that the current density remains within the range of $10^{-7}$ -$10^{-8}A/cm^2$ up to 4 MV/cm after BTA at 3 MV/cm, 550 K for $6x10^3$ s . No breakdown of the dielectric is seen up to 4 MV/cm. Figure 2(b) shows the comparison of the leakage current density between $Cu/SiO_2$ [7, 8] and $Cu/MnO/SiO_2$ under different BTA conditions. The figure indicates that the $Cu/SiO_2$ show very high leakage current density ($10^{-2}A/cm^{-2}$) under BTA at 2MV/cm, 423K. In contrast, the $Cu/MnO/SiO_2$ shows negligible increase in leakage current density. These results indicate good barrier properties of the MOCVD grown $MnO_x$ layer.

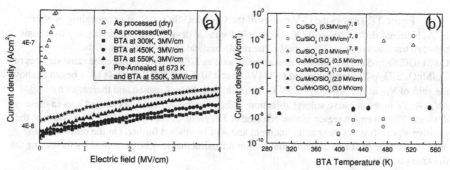

**Fig. 2** a) J-E curve of $Cu/MnO_x/SiO_2$/p-Si MOS devices; b) Leakage current density with (this work) and without (Refs. 7, 8) the $MnO_x$ barrier layer under different BTA conditions.

The capacitance-voltage (C-V) curves of the $Cu/MnO/SiO_2$/p-Si MOS devices are shown in Figs. 3 (a) and (b) to investigate the effects of annealing and BTA, respectively. The interface trap charge density ($D_{it}$) was found in the range of $10^{11}eV^{-1}cm^{-2}$ as determined from C-V analysis. The theoretical C-V curve of these MOS device is also plotted by creating appropriate digitize grid (ATHENA®) and solving them using Newton's methods in ATLAS. This is to obtain reference $V_{FB}$ values of $Cu/MnO/SiO_2$/p-Si MOS structure that has the interface trap charge density of $D_{it} \sim 10^{11}eV^{-1}cm^{-2}$. No hysteresis was observed between trace and retrace C-V curves for both the as-processed sample and the annealed samples, confirming no mobile charges in these devices. The value of flat band voltage ($V_{FB}$) is -1.04V for the thermal SiO$_2$ and is -0.4V for the p-TEOS SiO$_2$. These values are much smaller than the large $V_{FB}$ shift of -6.7 V in the Cu/SiO$_2$/p-Si devices. The large $V_{FB}$ shift in the sample without the MnO layer is due to sputtering damage and the penetration of foreign particles into the SiO$_2$ [5, 7, 8]. After thermal annealing of the Cu/ MnO/SiO$_2$/p-Si at 573, 673 and 773K for 3.6x10$^3$ s, no change is observed in the $V_{FB}$ shift, indicating no Cu ion diffusion into the SiO$_2$.

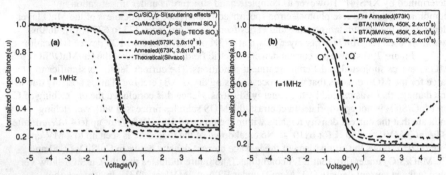

**Fig. 3.** C-V curves of Cu/MnO/SiO$_2$/p-Si MOS devices under different a) Processing conditions and b) BTA conditions.

Figure 3 (b) shows the C-V curves of the Cu/ MnO/SiO$_2$/p-Si after annealing or after BTA. Nominal shift of $\Delta V_{FB}$~0.1V is observed for these devices without having any hysteresis between trace and retrace C-V curves . The results indicate a good stability of the Cu/MnO/SiO$_2$/p-Si under thermal annealing as well as BTA. However once the same devices of Cu/MnO$_x$/SiO$_2$/p-Si were subjected to 3MV/cm at 450 K and 550 K for 2.4x10$^3$ s began to show the shift of $V_{FB}$ at first by $\Delta V_{FB}$~0.5V in the positive voltage direction and thereafter by $\Delta V_{FB}$~0.3V in the negative voltage direction, indicating the diffusion of charged ions in these devices. The inversion region is also affected at 3MV/cm, 550K for 2.4x10$^3$s. The origin of the positive $V_{FB}$ shift is not clear at the moment and will be studied further. On the other hand the negative voltage shift is due to nominal Cu ion diffusion that may be prevented by increasing the thickness of MnO.

## CONCLUSION

In conclusion, the ultrathin (~2nm) barrier layer of MnO was grown at 473K to examine the diffusion barrier property in Cu/MnO$_x$/SiO$_2$/p-Si MOS devices. The valence state of Mn is +2 and carbon content increases with $T_G$, however state of carbon in MnO$_x$ is under investigation. The Cu/ MnO/SiO$_2$/p-Si devices processed by wet etching showed very low leakage current density of 10$^{-8}$-10$^{-7}$A/cm$^2$ up to the applied filed of 4MV/cm as well as a low surface and interface trap charge density of 10$^{11}$eV$^{-1}$cm$^{-2}$. The current density remained within the same range before and after thermal annealing at 573, 673, and 773K for 3.6x10$^3$ s and after BTA at

3MV/cm at 550 K for $6 \times 10^3$ s. In addition, the C-V curves did not show any significant flat-band voltage shift, indicating negligible Cu ion diffusion in these devices. These properties make the MOCVD grown MnO a promising barrier material to for the multilayer interconnect structure in the advanced technology node.

## ACKNOWLEDGMENTS

One of the author VKD acknowledges Prof. Suto and Mr Hirota for technical help during the course of this study. This work was supported (in part) by Global COE Program "Materials Integration (International Center of Education and Research), Tohoku University, Japan and by Tokyo Electron Ltd. The author VKD is presently on EOL from Raja Ramanna Center for Advanced Technology, Indore, India.

## REFERENCES

1. P. Murarka, Metallization - Theory and Practice for VLSI-ULSI (Butterworth-Heineman, Boston,1993).
2. International Technology Roadmap for Semiconductors, Semiconductor Industry Association (2003).
3. K. Neishi, S. Aki, K. Matsumoto, H. Sato, H. Itoh, S. Hosaka, and J. Koike Appl. Phys. Lett., 93, 032106, (2008).
4. V. K. Dixit, K. Neishi, N. Akao and J. Koike (Unpublished).
5. S.P. Murarka, I. V. Verner, R. J. Gutmann, Copper - fundamental mechanisms for microelectronic applications, (Wiley, New York, 2000).
6. Takamasa Usui, Hayato Nasu, Shingo Takahashi, Noriyoshi Shimizu, T. Nishikawa, Masaki Yoshimaru, Hideki Shibata, Makoto Wada, and Junichi Koike. IEEE Transactions on Electron Devices, 53, 2492, (2006).
7. T. Suwwan de Felipe, Ph. D Thesis, Rensselaer Polytechnic Institute, Troy, NY, (1998).
8. T. Suwwan de Felipe, S. P. Murarka, S. Bedell, and W. A. Lanford, J. Vac. Sci. Technol. B, 15, 1987 (1997).

# Metallization II

Mater. Res. Soc. Symp. Proc. Vol. 1156 © 2009 Materials Research Society                    1156-D05-04

## Stress Gradients Observed in Cu Thin Films Induced by Capping Layers

Conal E. Murray[1], Paul R. Besser[2], Christian Witt[3], Jean L. Jordan-Sweet[1]

[1]IBM T.J. Watson Research Center, Yorktown Heights, NY

[2]GLOBALFOUNDRIES, Inc., Sunnyvale, CA

[3]GLOBALFOUNDRIES, Inc., T.J. Watson Research Center, Yorktown Heights, NY

## ABSTRACT

Glancing-incidence X-ray diffraction (GIXRD) has been applied to the investigation of depth-dependent stress distributions within electroplated Cu films due to overlying capping layers. 0.65 µm thick Cu films plated on conventional barrier and seed layers received a CVD $SiC_xN_yH_z$ cap, an electrolessly-deposited CoWP layer, or a CoWP layer followed by a $SiC_xN_yH_z$ cap. GIXRD and conventional X-ray diffraction measurements revealed that strain gradients were created in Cu films possessing a $SiC_xN_yH_z$ cap, where a greater in-plane tensile stress was generated near the film / cap interface. The constraint imposed by the $SiC_xN_yH_z$ layer during cooling from the cap deposition temperature led to an increase in the in-plane stress of approximately 180 MPa from the value measured in the bulk Cu. However, Cu films possessing a CoWP cap without a $SiC_xN_yH_z$ layer did not exhibit depth-dependent stress distributions. Because the CoWP capping deposition temperature was much lower than that employed in $SiC_xN_yH_z$ deposition, the Cu experienced elastic deformation during the capping process. Cross-sectional transmission electron microscopy indicated that the top surface of the Cu films exhibited extrusions near grain boundaries for the samples undergoing the thermal excursion during $SiC_xN_yH_z$ deposition. The conformal nature of these caps confirmed that the morphological changes of the Cu film surface occurred prior to capping and are a consequence of the thermal excursions associated with cap deposition.

## INTRODUCTION

The fabrication of Cu-based metallization in microelectronic technology involves numerous thermal excursions associated with the deposition and curing of constituent materials that comprise the back-end-of-line (BEOL). The thermal expansion mismatch between Cu and the underlying Si substrate can induce significant tensile stress in the metallization after these thermal cycles. Current manufacturing procedures employ electroplated Cu possessing Ta-based barrier layers that line the trench bottoms and sidewalls as well as capping layers on the top surface to limit interdiffusion between Cu and its environment and to mitigate electromigration along the metallization interfaces. Because tensile stress in the Cu features can facilitate the creation of voids, conditions that accentuate tensile stress must be properly understood and controlled. The presence of passivation layers above blanket Cu films has been shown to reduce stress relaxation at higher temperatures [1] by limiting diffusional mechanisms [2]. Since the interface between the Cu metallization and capping layers represents a location that is susceptible to electromigration-induced mass flow [3], a decrease in the relaxation of the tensile stress in this region represents a key reliability issue that must be investigated.

Grazing-incident x-ray diffraction (GIXRD) has been used to probe strain gradients within a variety of metallic films. By choosing the appropriate conditions for the incident and diffracted beam angles with respect to the film surface, the depth to which the diffraction information is collected can be controlled to dozens of nanometers near the critical angle [4]. A reduction of in-plane tensile stress was observed from GIXRD measurements on unpassivated Al(Cu) films [5, 6]. However, GIXRD characterization studies of Cu films with a free surface have reported either lower in-plane stress at the free surface [7], or a small increase (less than 40 MPa), suggested to be caused by oxidation at the surface [8]. Because the presence of surface oxidation can modify the mechanical behavior of Cu films, the current work focused on the effects of capping materials that act as diffusion barriers on the mechanical response of blanket Cu films, particularly near the Cu / capping layer interface.

## EXPERIMENTAL

30/10 nm thick Ta/TaN barrier layers and 80 nm thick Cu seed layers were sputter deposited onto 300 mm Si (001) substrates. Cu films were electroplated onto the Cu seed layers, followed by annealing at 100 $^0$C for 30 minutes in an inert atmosphere to promote grain growth. After chemical mechanical polishing, the final thickness of the Cu films was approximately 650 nm. Each wafer was then capped with a 35 nm thick $SiC_xN_yH_z$ film deposited at 350 $^0$C, an electrolessly deposited 7.5 nm CoWP film, or a 7.5 nm CoWP film followed by a 35 nm thick $SiC_xN_yH_z$ film deposited at 350$^0$ C. Although normal processing conditions produce an in-plane compressive stress in the $SiC_xN_yH_z$ film, the $SiC_xN_yH_z$ cap for one of the wafers was treated in a different manner that produced a residual tensile stress.

X-ray diffraction measurements were conducted at Brookhaven National Laboratory's National Synchrotron Light Source X20A beamline. A description of the experimental conditions can be found in [9]. A symmetric diffraction condition was chosen for the in-plane GIXRD measurements so that corrections due to refraction of the x-ray beam were not necessary [6,10]. Because the films possessed a large volume fraction of Cu (111) grains, the Cu (220) reflection, which corresponds to planes perpendicular to (111), was used ($2\theta \sim 68.7^0$). Measurements were conducted in reciprocal space (hkl), where diffraction peaks were collected along the in-plane reciprocal lattice vector h for various values of l, which ranged from 0.03 to 0.23, corresponding to incident and exit angles of 0.2$^0$ and 2.6$^0$, respectively. In glancing-incidence geometry, only the evanescent wave penetrates into the film below a critical angle, above which the penetration depth increases rapidly as the incidence angle is increased. Although the critical angle for a Cu film is approximately 0.38$^0$ for a beam energy of 8.6 keV, the presence of capping layers, which possess different scattering factors from those of Cu, modified this angle so that the exact penetration depth of the x-rays was different from that calculated for a single Cu film.

X-ray diffraction was also used to determine the bulk stress of the Cu films by measuring the Cu (222) interplanar spacing, d, as a function of $\psi$, the angle between the diffraction vector and the surface normal. Gaussian fits of the Cu (222) diffraction peak centers were converted into values of d using Bragg's law. An isotropic, biaxial in-plane stress state was assumed for the blanket films, so that the out-of-plane normal stress and all shear stresses were assumed to be zero. Equation 1 represents the link between the fitted d vs. $\sin^2(\psi)$ slope to the in-plane biaxial stress, $\sigma_{bulk}$, for a quasi-isotropic elastic material [11]

$$\frac{\partial d_{222}}{\partial \sin^2(\psi)} = d_0 \left(\frac{1+\nu}{E}\right)_{XEC} \sigma_{bulk} \qquad (1)$$

where the x-ray elastic constants (XEC), E and $\nu$ represent the effective Young's modulus and Poisson's ratio, respectively, of the diffracting grains. The use of the slope of the $d_{222}$ vs. $\sin^2(\psi)$ fitted line in Equation 1 for each sample rather than subtracting an unstressed lattice spacing, $d_0$, from the lattice spacing measurements minimizes the uncertainty associated with the actual $d_0$. This error is placed in the pre-factor, and is estimated to be less than 0.1%. $d_0$ was assumed to be the value of the intercept of the fitted $d_{222}$ line ($\psi = 0^0$). The XEC on the right hand side of Equation 1 were calculated using the Neerfeld-Hill limit, which is an average of the Voigt and Reuss values [11], where for Cu (222), $(1+\nu)/E = 7.912 \times 10^{-3}$ GPa$^{-1}$. Although Cu (111) represented the dominant texture component of the 3 samples, the measured Cu (222) interplanar spacings taken at several $\psi$-tilts all revealed linear behavior with respect to $\sin^2(\psi)$.

## RESULTS

Figure 1a depicts the corresponding d-spacing of the Cu (220) peak measured in grazing-incidence as a function of the incident beam angle above the surface of the $SiC_xN_yH_z$ capped Cu film. A gradient in the Cu lattice spacing is evident at small incidence angles, where a larger in-plane lattice spacing is observed near the top surface of the Cu film relative to that in the bulk. Because GIXRD measurements represent a volume average through the penetration depth of the X-rays, the depth of the diffracting volume increases rapidly as the incident angle is raised above the critical angle. However, the measured Cu (220) spacing for the sample with a CoWP capping layer only, as shown in Figure 1b, exhibits no gradient in lattice spacing through the film.

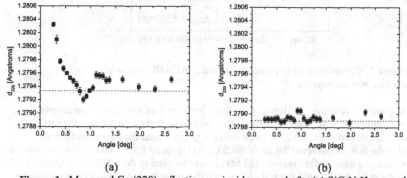

(a)            (b)

**Figure 1.** Measured Cu (220) reflection vs. incidence angle for (a) $SiC_xN_yH_z$-capped Cu film and (b) CoWP-capped film. Dotted lines refer to extrapolated in-plane bulk values for the $SiC_xN_yH_z$-capped (1.2793 Å) and CoWP-capped Cu films (1.2789 Å), respectively.

For all samples which possessed a $SiC_xN_yH_z$ capping layer, the Cu (220) in-plane spacing was larger near the top surface of the film. The interplanar spacing decreased as the incidence angle was increased, eventually reaching the average value for the bulk film. However, the Cu lattice spacings corresponding to the bulk of the films varied due to the difference in processing conditions among the samples. To calculate the increase in in-plane normal stress associated with $d_{220}^{surf}$, the near-surface lattice spacings measured using GIXRD, we use Equation 2:

$$\sigma_{surf} = \sigma_{bulk} + \left(\frac{d_{220}^{surf} - d_{220}^{bulk}}{d_{220}^{bulk}}\right)\left(\frac{1-\nu}{E}\right)_{XEC} \quad (2)$$

where $d_{220}^{bulk}$ corresponds to the value of the lattice spacing as extrapolated from the fit of the Cu (222) measurements to the in-plane orientation ($\psi = 90^0$). The dotted red lines in Figures 1a and 1b represent this extrapolated lattice spacing, where the values were normalized by the ratio of the Miller indices: $d_{220} = d_{222}\sqrt{12/8}$. Since the grazing-incidence measurements of the (220) reflection primarily occur from the Cu (111) textured grains, the XEC for Equation 2 were calculated in the Neerfeld-Hill limit using the (222) reflection: $(1-\nu)/E = 4.238 \times 10^{-3}$ GPa$^{-1}$.

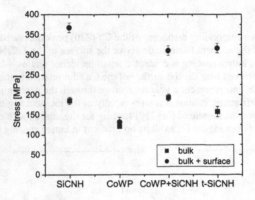

**Figure 2**. Comparison of in-plane stress values in Cu films in bulk and near-surface regions.

The in-plane stress of the Cu films as calculated from the d vs. $\sin^2(\psi)$ measurements (Equation 1), corresponding to the bulk film value, and those in the near-surface regions (Equation 2) are shown in Figure 2. A large increase in stress near the top Cu surface, as compared to the bulk, is observed for all of the SiC$_x$N$_y$H$_z$-capped films. The greatest increase in tensile stress in the near-surface region (182 MPa) was observed in the SiC$_x$N$_y$H$_z$-capped Cu film. As the results in Figure 2 would suggest, the sample that possessed only a CoWP cap exhibited a negligible increase in tensile stress near the surface. However, for the Cu film that possessed both a CoWP and SiC$_x$N$_y$H$_z$ cap, the increase in tensile stress was approximately 65% of that in the other SiC$_x$N$_y$H$_z$ capping case (117 MPa). The stress at the Cu / cap interface for the sample with a tensile SiC$_x$N$_y$H$_z$ cap (t-SiCNH) also exhibited an increase of approximately 159 MPa.

Cross-sectional transmission electron microscopy (XTEM) was also performed on these samples to investigate the microstructure of the Cu films. Figure 3 depicts an XTEM image of the Cu film with a conventional SiC$_x$N$_y$H$_z$ cap. Shallow extrusions were found in all samples

cap →

Cu

Ta/TaN →
Si

200 nm

**Figure 3**. Cross-sectional TEM image of SiC$_x$N$_y$H$_z$-capped Cu film

that received SiC$_x$N$_y$H$_z$ capping. However, the SiC$_x$N$_y$H$_z$ caps followed the underlying morphology of the top Cu surfaces. No extrusions were observed in the sample with only an electrolessly deposited CoWP cap.

## DISCUSSION

The results indicate that the thermal excursion during capping, in addition to the constraint imposed by the overlying capping layer, drive the increase in tensile stress near the film / cap interface. The presence of passivation layers on the top surface of Cu films have been shown to reduce stress relaxation at temperatures above 300 $^0$C [1,2]. Also, both CoWP and SiC$_x$N$_y$H$_z$ caps decrease Cu diffusion at this interface [12]. However, Cu films only experience a small temperature increase during CoWP deposition (less than 100 $^0$C), resulting in predominantly elastic deformation. The SiC$_x$N$_y$H$_z$ deposition process involves temperatures (~350 $^0$C) at which significant plastic deformation can occur to partially relieve the compressive stress generated in the Cu films due to the coefficient of thermal expansion (CTE) mismatch from the substrate. The extrusions, which were observed at grain boundaries as indicated in Figure 3, reveal significant Cu diffusion to reduce the overall stress state. The conformal nature of the SiC$_x$N$_y$H$_z$ cap over these extrusions confirms that the diffusion occurred prior to capping. The plastically-deformed microstructure is stabilized by the presence of the cap which limits subsequent Cu diffusion near the cap / Cu interface as the sample is cooled. Upon cooling from the deposition temperature, the CTE mismatch between Cu and the substrate and SiC$_x$N$_y$H$_z$ cap drives the Cu film into a sufficient value of tension to activate plastic relaxation. In fact, in the absence of plasticity, the Cu film stress would theoretically approach 1 GPa, indicating that relaxation is occurring throughout the depth of the film. As seen in Figure 2, the additional thermal cycle that the Cu films experience during SiC$_x$N$_y$H$_z$ capping increases the bulk Cu stress (190 MPa) relative to the CoWP-only capped Cu film (120 MPa). The constraint imposed by the strong adhesion between the Cu and SiC$_x$N$_y$H$_z$ cap leads to an even greater tensile stress near the Cu / SiC$_x$N$_y$H$_z$ interface.

In addition, the Cu film possessing both CoWP and SiC$_x$N$_y$H$_z$ caps exhibits an increase in the in-plane, near-surface Cu stress of 117 MPa as compared to 182 MPa for the Cu film with the conventional SiC$_x$N$_y$H$_z$ cap. This effect may be explained by the presence of the CoWP cap, which can act as a buffer layer between the SiC$_x$N$_y$H$_z$ and the Cu film. Although the same increase in tensile stress may exist near the CoWP / $SiC_xN_yH_z$ interface as was observed near the Cu / $SiC_xN_yH_z$ interface, the amorphous nature of the CoWP film does not allow us to determine its strain using diffraction-based methods. Also, residual stress present in the $SiC_xN_yH_z$ cap plays a smaller role on the increase in stress near the surface of blanket Cu films. The Cu film with a residual tensile $SiC_xN_yH_z$ cap (t-SiCNH) exhibited an increase of 159 MPa, as compared to 182 MPa in the case of a compressively stressed $SiC_xN_yH_z$ capping layer. The net force of the capping layer is primarily accommodated by the underlying Si substrate in the case of blanket samples. However, in the case of patterned Cu features, the overlying cap may play a greater role on the stress distributions generated in the Cu and must be investigated.

## CONCLUSIONS

The in-plane stress near the top surface of Cu films exhibits a larger value than that of the bulk Cu for samples possessing a $SiC_xN_yH_z$ capping layer. However, samples with a CoWP cap, which did not receive the same thermal treatment, experienced elastic deformation and did not show a gradient in the Cu stress. The $SiC_xN_yH_z$ capping layer constrains the near-surface region of the Cu film, thereby limiting the extent of plastic deformation during cooling after $SiC_xN_yH_z$ deposition. Extrusions observed at the top surface of Cu films that received $SiC_xN_yH_z$ caps confirmed that significant plastic relaxation of the compressive stress induced by the CTE mismatch occurs prior to cap deposition. However, the increase in tensile stress at the near-surface regions of the Cu films represents an important reliability issue because of its susceptibility to stress-voiding phenomena.

## ACKNOWLEDGMENTS

This work was performed by the Research Alliance Teams at various IBM Research and Development facilities. The diffraction measurements were carried out at the National Synchrotron Light Source, Brookhaven National Laboratory, which is supported by the U.S. Department of Energy, Division of Materials Sciences and Division of Chemical Sciences, under Contract No. DE-AC02-98CH10886.

## REFERENCES

1. R.P. Vinci, E.M. Zielinski and J.C. Bravman, Thin Solid Films, **262**, 142 (1995).
2. R-M. Keller, S.P. Baker, E. Arzt, J. Mater. Res. **13**, 1307 (1998).
3. C.K. Hu, R. Rosenberg, K.Y. Lee, Appl. Phys. Lett. **74**, (1999).
4. L.G. Parratt, Phys. Rev. **95**, 359 (1954).
5. M.F. Doerner and S. Brennan, J. Appl. Phys. **63**, 126 (1988).
6. C.J. Schute and J.B. Cohen, J. Appl. Phys. **70**, 2104 (1991).
7. T. Himuro and S. Takayama, MRS Symp. Proc. **854**, U11.11.1 (2005).
8. S. Takayama, M. Oikawa and T. Himuro, MRS Symp. Proc. **795**, U5.11.1 (2004).
9. C.E. Murray, P.R. Besser, C. Witt, J.L. Jordan-Sweet, Appl. Phys. Lett. **93**, 221901 (2008).
10. M.F. Toney and S. Brennan, Phys. Rev. B, **39**, 7963 (1989).
11. I.C. Noyan and J.B. Cohen, *Residual Stress*, (Springer-Verlag, NY, 1987).
12. D. Gan, P.S. Ho, Y. Pang, R. Huang, J. Leu, J. Maiz and T. Scherban, J. Mater. Res. **21**, 1512 (2006).

# Reliability

Mater. Res. Soc. Symp. Proc. Vol. 1156 © 2009 Materials Research Society 1156-D06-06

# Large-Scale Electromigration Statistics for Cu Interconnects

Meike Hauschildt, Martin Gall, *Richard Hernandez

Silicon Technology Solutions, Freescale Semiconductor, Inc.
2070 Route 52, Hopewell Junction, NY 12533, USA
*3501 Ed Bluestein Blvd, MD K10, Austin, TX 78721, USA

## ABSTRACT

Even after the successful introduction of Cu-based metallization, the electromigration failure risk has remained one of the important reliability concerns for advanced process technologies. The observation of strong bimodality for the electron up-flow direction in dual-inlaid Cu interconnects has added complexity, but is now widely accepted. More recently, bimodality has been reported also in down-flow electromigration, leading to very short lifetimes due to small, slit-shaped voids under vias. For a more thorough investigation of these early failure phenomena, specific test structures were designed based on the Wheatstone Bridge technique. The use of these structures enabled an increase of the tested sample size past 1.1 million, allowing a direct analysis of electromigration failure mechanisms at the single-digit ppm regime. Results indicate that down-flow electromigration exhibits bimodality at very small percentage levels, not readily identifiable with standard testing methods. The activation energy for the down-flow early failure mechanism was determined to be 0.83 ± 0.01 eV. Within the small error bounds of this large-scale statistical experiment, this value is deemed to be significantly lower than the usually reported activation energy of 0.90 eV for electromigration-induced diffusion along Cu/SiCN interfaces. Due to the advantages of the Wheatstone Bridge technique, we were also able to expand the experimental temperature range down to 150 °C, coming quite close to typical operating conditions up to 125 °C. As a result of the lowered activation energy, we conclude that the down-flow early failure mode may control the chip lifetime at operating conditions. The slit-like character of the early failure void morphology also raises concerns about the validity of the Blech-effect for this mechanism. A very small amount of Cu depletion may cause failure even before a stress gradient is established. We therefore conducted large-scale statistical experiments close to the critical current density-length product *(jL)\**. The results indicate that even at very small failure percentages, this critical product seems to extrapolate to about 2900 ± 400 A/cm for SiCOH-based dielectrics, close to previously determined *(jL)\** products of about 3000 ± 500 A/cm for the same technology node and dielectric material, acquired with single link interconnects. More detailed studies are currently ongoing to verify the extrapolation methods at small percentages. Furthermore, the scaling behavior of the early failure population was investigated.

## INTRODUCTION

The continuing drive towards smaller interconnect circuitry as well as the introduction of new materials and deposition processes for dielectric and barrier applications results in reliability concerns due to electromigration (EM). EM failures in Cu interconnects are mainly the result of void growth at the cathode end of the line. The corresponding times to failure (TTF) usually follow a lognormal distribution with the median lifetime (MTTF) defined by the quality of the top interface, which controls the mass transport [1-4]. Since EM experiments are performed at higher current densities and temperatures compared to operating conditions, extrapolations are necessary to assess reliability at operating conditions using the following equation

$$TTF_{oper} = MTTF_{stress}\left(\frac{I_{stress}}{I_{oper}}\right)^{n} \exp\left[\frac{E_a}{k_B}\left(\frac{1}{T_{oper}} - \frac{1}{T_{stress}}\right) + NSD * \sigma\right] \tag{1}$$

where $I$ is the current, $n$ the current exponent, $E_a$ the activation energy, $k_B$ Boltzmann's constant, $T$ the temperature, $NSD$ the number of standard deviations, and $\sigma$ the lognormal standard deviation. The subscripts 'stress' and 'oper' refer to EM testing and actual operating conditions, respectively. Equation 1 indicates that the main factors requiring attention are the activation energy related to the dominating diffusion mechanism, the current exponent as well as the median lifetimes and lognormal standard deviation values of experimentally acquired failure time distributions. Whereas the origins of EM activation energy and current exponent as well as the behavior of median lifetimes with continuing device scaling are relatively well understood [e.g. 3, 5], the origin and scaling behavior of the lognormal standard deviation has only recently been studied in detail [6-10]. Adding to the complexity of the standard deviation is strong bimodality for the electron up-flow direction in dual-inlaid Cu interconnects [11-13]. The failure voids can occur both within the via ("early" mode) or within the trench ("late" mode). More recently, bimodality has been reported also in down-flow EM, leading to very short lifetimes due to small, slit-shaped voids under vias [14, 15]. In addition to shorter MTTFs, these early failure modes can have different activation energy, current exponent and sigma values compared to the late mode depending on the diffusion and void formation mechanisms. To evaluate these early failure phenomena, specific test structures were designed based on the Wheatstone Bridge technique. Reference [16] contains a detailed characterization of activation energy and current exponent values. The objective of this study is the continuation of that thorough investigation. This study will focus on an extension of the large sample size achieved earlier. In addition, the well-known Blech effect was examined employing Wheatstone Bridges, extending its applicability towards small failure percentages. To conclude this study, technology scaling was investigated using samples from the 65nm process node, and compared with the predominantly used 90nm samples. The scaling behavior was found to follow the general trend of shorter lifetimes as a consequence of reduced interconnect height and via size.

## EXPERIMENTAL PROCEDURE: STATISTICAL EM TESTING METHODOLOGY

Standard EM testing is performed on a Cu line connected by vias on both ends of the structure to a lower or upper metal level. The current flows from a wide metal line through the via into the test line. If the test line is on a lower metal level compared to the current supply line, the test is called 'down-flow', whereas it is called 'up-flow' for test lines on a higher metal level. Only one via/line interface is being tested per sample. If only a few percent of these single line test structures belong to the early failure population, as generally observed especially in down-flow EM, the identification of separate statistics for the early and late modes is not possible. Even a few hundred or thousands of samples might not be sufficient to extract the necessary parameters. However, for an extrapolation to use conditions, the assumption of a monomodal distribution might not be adequate. Since it is not feasible to test enough parts using standard, single link structures, a statistical testing methodology based on the well-known Wheatstone Bridge technique was used for a detailed examination of these failures.

Several studies employed statistical test structure designs to examine the void formation behavior in interconnects [3, 11, 15-20]. The test structures used in this study consist of a Wheatstone Bridge arrangement of parallel and serial interconnects. Current flows through all of the lines simultaneously and the resistance imbalance of the Wheatstone Bridge is being monitored over time. A change in the imbalance indicates failure void formation in one of the test lines. This particular test design was successfully employed for early failure studies in Al(Cu) interconnects [17, 18] and recently for Cu metallization [15, 16, 20]. Details can be found in these references and specifically in [16]. The resistance change resolution and therefore the sensitivity to detect EM induced void formation is greatly increased by the Wheatstone Bridge technique, basically enabling an increase in the number of tested links per device without the loss of detection of small resistance changes. In this study, three different test structure designs were used. The 90nm technology samples consisted of either 920 or 760 interconnects each. The line width is 0.12 μm for down-flow and 0.14 μm for up-flow EM, respectively. The line length was 50 μm or 70 μm instead of 250 μm for standard EM structures, a result of test system limitations. However, at the reference current density of 1.5 MA/cm$^2$ used in this study no differences were observed between 50 μm and 250 μm long standard structures validating a direct comparison between the shorter statistical structures and the 250 μm single link samples. The 65nm technology test structures contained 2448 interconnects instead of just one in standard test designs with a line width of 0.09 μm and a length of 250 μm. The Cu interconnects are surrounded by a Ta-based barrier layer and a SiCN cap layer. SiCOH is used as the dielectric at metal and via levels.

Down-flow EM experiments were performed at seven temperatures between 150 °C and 325 °C and current densities in the range of 0.15 and 1.5 MA/cm$^2$. Up-flow tests were limited to 325 °C and 1.5 MA/cm$^2$. 126 up-flow and 677 down-flow Wheatstone Bridge devices were tested for the 90nm technology resulting in a total number of examined interconnects of 115,920 and 686,400 for up-flow and down-flow directions, respectively. 306,000 down-flow interconnects were tested using 65nm processing, increasing the total sample size past 1.1 million.

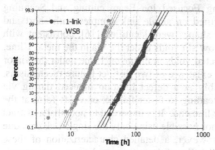 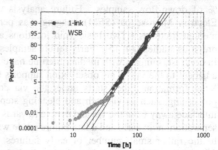

**FIGURE 1.** Lifetime distributions for down-flow Wheatstone Bridge devices and the corresponding single link samples at 325 °C.

**FIGURE 2.** Deconvoluted lifetime data for down-flow Wheatstone Bridge devices and the corresponding single link data at 325°C.

Figure 1 represents a plot of lifetime distributions for down-flow Wheatstone Bridge devices and the corresponding single link samples tested at 325 °C and 1.5 MA/cm$^2$. The lines through the failure data using a monomodal lognormal fit are for illustrative purposes only. A significant decrease in lifetimes can be discerned for the Wheatstone samples compared to standard single link devices. This is in agreement with the weakest link approach, i.e. the failure of the weakest link determines the lifetime of the whole assembly of multiple links. It needs to be remembered that one data point in the Wheatstone Bridge distribution represents the first failed line out of 920 interconnects. Thus, while showing only 134 data points, the Wheatstone distribution contains information about 123280 interconnects. However, once the first link out of 920 interconnects within a Wheatstone Bridge device has failed, generally no detailed information about the remaining 919 links can be extracted, since all other links start to fail consecutively and contribute to the total resistance imbalance change.

In order to assess the EM early failure mechanism, the Wheatstone Bridge failure data shown in Figure 1 needs to be represented with respect to the single interconnect level, i.e. the limited information about the non-failed links in the EM failure distribution needs to be included to enable a direct comparison between Wheatstone and single link structures. The statistical deconvolution is accomplished using conditional reliabilities in conjunction with censored data analysis. As a result, each failure time is newly associated with a cumulative failure percentage. Cumulative distribution plots can now be plotted using either the failure percentage or the corresponding number of standard deviations (NSD) as the ordinate. For plotting convenience, most graphs display the latter. Details of this calculation can be found in reference [16].

## BIMODALITY IN DOWN-FLOW EM

Deconvoluted lifetime data for down-flow Wheatstone Bridges tested at 325 °C and 1.5 MA/cm$^2$ is shown in Figure 2 together with single link data. Note that the lifetimes are identical to the ones depicted in Figure 1, only the probabilities are different. The deconvoluted Wheatstone Bridge data and the single link results align well at the transition point. However, it is evident that the majority of the Wheatstone Bridge data deviates from the monomodal behavior which is represented by the straight line fit through the single link data. This observation clearly indicates the existence of an early failure mechanism for approximately 0.1 % of down-flow samples. Failure analysis using Focused Ion Beam cutting and Scanning Electron Microscope (SEM) imaging was performed on single link samples with short and median lifetimes. Two basic void locations and shapes were identified [16], consistent with previous studies [14, 15]. Early failure samples show slit voids under the via in the metal 1 line. In contrast, samples with median lifetime have faceted voids in the metal 1 line under the via or adjacent to the via. This observation supports the existence of two distinct failure mechanisms. Void formation either starts under the via possibly at a defect location induced during via processing, such as via etch and cleaning steps, leading to faster failure, or voids form at the interface away from the via and evolve towards the line end. It is important to note that a distinct separation of an early and a late failure mode was not possible using only the single link data due to the rather small number of early failures. However, a detailed characterization of these failures is essential, since these devices limit the reliability of the entire population.

Lifetime distributions of down-flow Wheatstone Bridge devices corresponding to seven temperatures in the range of 150 °C to 325 °C are shown in Figure 3. All experiments were performed at 1.5 MA/cm$^2$. The data in this figure is not deconvoluted, i.e. one data point represents 920 interconnects. The lifetime distributions show consistent MTTF and sigma values, with increasing MTTF as the temperature decreases and comparable sigma values over the entire temperature range. Figure 4 displays the deconvoluted Wheatstone Bridge data as well as the corresponding single link results. The distributions originating from multi link and single link testing align reasonably well at all temperatures with only slight disconnects at 275 °C and 250 °C. Due to the long testing times, no single link data at 175 °C and 150 °C is available. The Wheatstone Bridge distributions show consistently higher sigma values indicating the existence of the early failure mode at all temperatures. According to the weakest link approach, it is expected that the majority of the multi link samples fail according to the early failure mechanism. Thus, they are thought to be caused by slit-like failure voids under the via. It appears that the percentage of these fails is small. However, it needs thorough investigation since this failure population defines the actual chip lifetime.

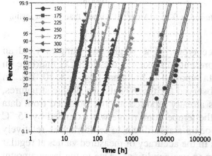

**FIGURE 3.** Lifetime distributions of down-flow Wheatstone Bridge devices as function of temperature.

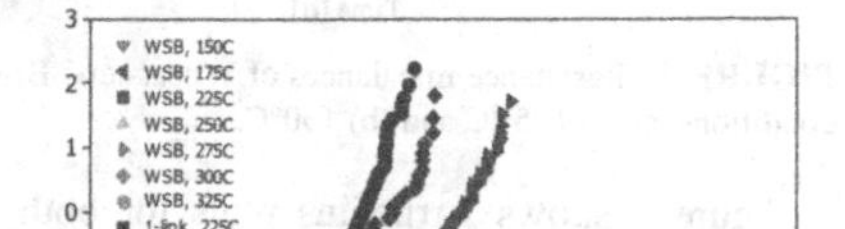

**FIGURE 4.** Deconvoluted lifetime data for down-flow Wheatstone Bridge devices and single link data.

A few very early fails were found at 175 °C and 150 °C possibly indicating a "third" failure mode. It has been observed that stress-induced voiding in Cu interconnects occurs predominantly in the 150 °C to 200 °C range [e.g. 21]. To examine whether stress-induced voiding could have caused the observed very early failures under EM testing conditions, Wheatstone Bridge testing was performed at 0.15 MA/cm$^2$, 10x lower than regular testing conditions to rule out any EM-induced effects. 13 and 9 samples were tested at 175 °C and at 150°C, respectively, resulting in 11960 and 8280 examined interconnects. Resistance imbalances as a function of time for both temperatures are shown in Figure 5. No failure was observed in more than 1.9 years of testing at 175 °C and 2.3 years of testing at 150°C. As the very early fails in the EM tests occurred significantly earlier, "classical" stress-induced voiding is unlikely to be the cause of these fails. However, "stress-assisted" EM is a possibility as the Cu lines are subject to higher tensile stresses at these testing temperatures, possibly reducing the critical stress level for EM void formation and/or facilitating EM mass transport.

Ideally, to determine activation energy values for both failure modes, a bimodal fit to the experimental lifetime distribution needs to be obtained. It is reasonable to assume that both mechanisms follow lognormal distribution characteristics, since their kinetics are both controlled by the top interface of the metal 1 line. However, due to the large size of the data set, this is a rather challenging task. Since the majority of the multi link devices appear to fail according to the early mode and most of the single link samples display late mode characteristics, it seems to be a reasonable approximation to determine the characteristics of the early mode from the Wheatstone Bridge devices and the parameters of the late mode from the single link samples.

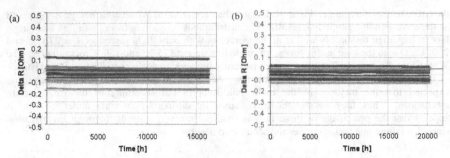

**FIGURE 5.** Resistance imbalances of Wheatstone Bridges tested at 0.15 MA/cm$^2$, 10x lower than regular testing conditions, at (a) 175 °C and (b) 150°C.

Figure 6 shows Arrhenius plots for both device types. The single link samples yield an activation energy of 0.91 ± 0.03 eV. This value represents diffusion at the Cu/SiCN interface on top of the metal 1 line, in good agreement with data reported in literature [e.g. 3, 5, 10, 19]. In contrast, a slight decrease in activation energy is observed for the Wheatstone Bridge structures, namely a value of 0.83 ± 0.01 eV. This value was already reported in [16], obtained from data taken between 325 °C and 225 °C. In this study, the temperature range was extended by 75 °C down to 150 °C. Thus, the activation energy appears stable down to temperatures relatively close to use conditions. This fact raises confidence in the accuracy of measured values at regular accelerated EM test conditions for the extrapolation to use conditions. Due to the large amount of data and resulting small error bars, the difference between early and late failure populations is deemed to be significant. Thus, the early failure mode in down-flow samples shows a decrease in activation energy. This change is possibly the result of void formation being defined by the quality of the interface between via bottom and metal line, and not solely by the diffusion along the top metal 1 interface. In addition, the elevated tensile stress level in the Cu line (directly under the via) may lead to a reduction in the apparent activation energy for the early failure mechanism.

## COMPARISON BETWEEN DOWN-FLOW AND UP-FLOW EM DATA

Figure 7 shows the lifetime data for down-flow as well as up-flow Wheatstone bridge devices and single link structures at 325 °C and 1.5 MA/cm$^2$. The deconvoluted up-flow Wheatstone Bridge and single link data align well at the transition point. Thus, the multi link structures confirm the bimodality in the up-flow failure distribution already clearly observable with single link testing. The TTF values of the early mode in up-flow samples originating from Wheatstone samples as well as single link structures are significantly smaller compared to the early mode data in down-flow devices. Furthermore, it seems that the up-flow samples show a third failure mode exhibiting very long lifetimes possibly due to barrier leakage. However, the regular late mode failures in up- and down-flow samples appear to show comparable MTTF and sigma values. These results indicate that at accelerated testing conditions the up-flow interface is more critical than the down-flow direction in defining the chip lifetime. From these accelerated testing conditions the lifetimes need to be extrapolated to use conditions employing Equation 1. The target operating conditions, namely temperature and current density, as well as the failure rate are defined differently by each company depending on the customer needs. While generally all of these parameters need to be considered, in the following only the temperature was taken into account when extrapolating from stress conditions to operating conditions. Since the Wheatstone data provides actual lifetime values for low ppm levels, such as 1000, 100 and 10 ppm, extrapolations can be performed using these measured values instead of employing the MTTF and sigma values of the respective early modes. Thus, Equation 1 simplifies to

$$TTF_{oper} = TTF_{stress} \exp\left[\frac{E_a}{k_B}\left(\frac{1}{T_{oper}} - \frac{1}{T_{stress}}\right)\right] \quad (2)$$

where $TTF_{stress}$ corresponds to the measured

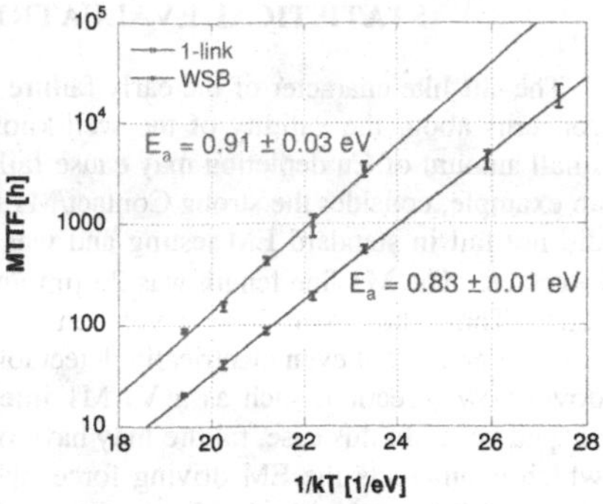

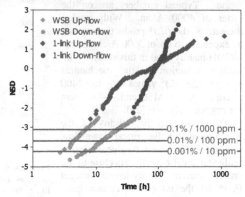

**FIGURE 6.** Arrhenius plots for down-flow Wheatstone bridge structures and standard single link samples indicating different activation energies.

failure time at a chosen cumulative percent. While it is important to examine the extrapolated values of all failure modes, the early mode is expected to be the limiting factor in determining chip lifetime. Thus, the activation energy values determined for the early failure modes were used for the calculations here. As reported in [16], an activation energy value of 0.93 ± 0.03 eV was obtained for the early failure mode of up-flow tested samples. The early population in down-flow EM seemed to show a value of 0.83 ± 0.01 eV. The chosen stress and operating temperatures were 325 °C and 105 °C. Performing the extrapolation at a 1000 ppm level, the early mode for the up-flow direction yields a smaller lifetime compared to down-flow EM in agreement with the trend of the measured values. However, at 10ppm and at 105 °C, down-flow EM appears to yield smaller lifetimes than up-flow EM as a result of the smaller activation energy value, namely 8.7 years vs. 12.4 years. Overall, all extrapolated values are rather high providing passing data for typical reliability specifications. However, it seems possible that a different interface determines the chip lifetime at accelerated testing conditions and selected use conditions. Further extrapolations might be necessary depending on each company's individual current target using experimentally determined values for the current exponent. In general, this result clearly shows the importance of a thorough characterization of all possible failure modes, especially at the small percentage level. Particularly the determination of activation energy and current exponent is needed, but depending on the failure rate target, the evaluation of MTTF and sigma values is also important.

**FIGURE 7.** Lifetime distributions for up-flow and down-flow Wheatstone Bridges and single link devices at 325 °C and 1.5 MA/cm².

## STATISTICAL EVALUATION OF SHORT-LENGTH EFFECTS

The slit-like character of the early failure void morphology in the EM down-flow case raises concerns about the validity of the well-known Blech-effect for this mechanism [22]. A very small amount of Cu depletion may cause failure even before a stress gradient is established. As an example, consider the strong Contact/M1 up-flow interface shown in Figure 8. This structure did not fail in standard EM testing and was part of a Blech-effect study using the Contact/M1 interface. The M1 line length was 25 μm and the resistance increase saturated well below 10% [23]. Thus, the small slit-like void on top of the M1 line would not lead to any reliability concern and is not even electrically detectable; however, it would lead to a complete open in the down-flow direction, such as a V1/M1 interface, if the via were located right above the slit-shaped void. In this case, failure may have occurred well before the build-up of a stress gradient which counteracts the EM driving force, whereas enough material was depleted to generate a stress gradient in the M1 line for the Contact/M1 case.

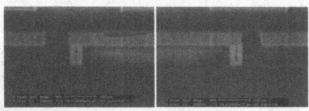

**FIGURE 8.** Slit-shaped void at the top of an M1 line, connected to contacts. This interconnect did not fail; however, the same void could have lead to failure for the V1/M1 down-flow case.

The Blech effect has been studied to quite some extent over the last 30+ years. For its application in Cu metallization, see, e.g. [23-25]. Figure 9 summarizes some of the results. For mechanically strong dielectrics, such as TEOS and F-TEOS, relatively high critical current density-length products $(jL)^*$ have been found. Typical numbers are on the order of 4000 A/cm. With weaker dielectrics, the $(jL)^*$ products decrease, as expected, to about 3000 A/cm for the SiCOH mainly used in this study. With further reduction in the mechanical strength, the $(jL)^*$ values drop to ~2400 A/cm for porous MSQ and ~1400 A/cm for organic polymers [25].

To investigate the validity of the Blech-effect for the critical down-flow EM mechanism, we conducted large-scale statistical experiments close to the critical current density-length product $(jL)^*$ for the integration chosen here. The interconnect line length was 50 μm. Based on single interconnect EM

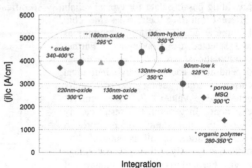

**FIGURE 9.** Trend of critical current density-length products, $(jL)^*$, as a function of dielectric. With mechanically weaker films, the values decrease. The encircled area encompasses TEOS/FTEOS-based integrations. The 130nm-hybrid integration uses FTEOS at the via level and SiCOH at the trench level (red datapoints: [23], *[25], **[24]).

128

experiments, the critical product for the strong Contact/M1 interface was determined earlier to be 3000 ± 500 A/cm [23]. Multiple current density-dependent experiments were run with Wheatstone Bridges in the range of 0.7 to 1.2 MA/cm$^2$ and compared to the standard current density condition of 1.5 MA/cm$^2$. The failure distributions are shown in Figure 10. Even though the lowest current density of 0.7 MA/cm$^2$ is very close to the expected critical current density of 0.6 MA/cm$^2$ (using single link Contact/M1 data), there are no

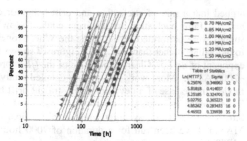

**FIGURE 10.** Wheatstone Bridge lifetime distributions as function of current density reaching close to the expected critical current density of 0.6 MA/cm$^2$.

indications of an increase in the lognormal sigma of the failure distribution, as is usually found with single interconnect testing [e.g. 23, 26]. In order to extract the critical $(jL)^*$ product, the inverse median lifetimes were plotted as a function of $(jL)$ in Figure 11. The Wheatstone Bridge structures were run at both 275 °C and 325 °C and linear extrapolations result in the same critical product of 2900 ± 400 A/cm. The single link Contact/M1 data from reference [23] are also shown. As mentioned above, this data extrapolates to 3000 ± 500 A/cm. These results indicate that even at very small failure percentages, the critical product seems to remain stable at about 3000 A/cm for SiCOH-based dielectrics.

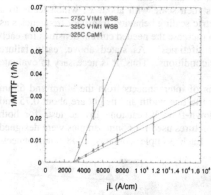

**FIGURE 11.** Inverse median failure times as function of $(jL)$. Linear extrapolations result in the same critical current density-length product of about 3000 A/cm.

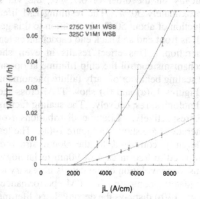

**FIGURE 12.** Inverse median failure times as function of $(jL)$, as shown in Figure 16. The extrapolation is performed using a current density exponent of $n=1.5$ instead of $n=1$ resulting in $(jL)^*$ of about 2000 A/cm.

Questions about the validity of the extrapolation methods to the critical $(jL)^*$ product remain. Usually, it is assumed that the inverse of the EM median lifetimes is proportional to $(jL)$ and that the critical intercept can be determined by a linear extrapolation towards zero effective drift, i.e. infinite lifetimes. The underlying current density exponent is therefore assumed to be close to $n=1$. In the case of a deviation from $n=1$, the extrapolation is not as straightforward. It has been

129

argued that a mix of nucleation and drift behavior could increase the current density towards $n=1.5$. Pure nucleation mechanisms could follow $n=2$ [27-29]. Current density exponent studies were performed earlier using the Wheatstone Bridge technique described here and have found values higher than $n=1$ [16]. Figure 17 shows an extrapolation assuming a current density exponent of $n=1.5$. Clearly, the value of the $(jL)^*$ product would decrease considerably towards ~2000 A/cm. More work and a more detailed analysis of the applicability of the extrapolation models in the case of a deviation from straight drift behavior, characterized by $n=1$, are needed. Experiments are currently ongoing with current density values at the expected critical level of 0.6 MA/cm$^2$, as well as slightly below this critical level at 0.5 MA/cm$^2$. Experiments at 0.35 MA/cm$^2$ have been run for a duration of 7000 hours without failures, giving a lower limit of ~1750 A/cm for the $(jL)^*$ product.

## TECHNOLOGY SCALING

With continuing scaling of interconnects, EM lifetimes in single link tests were found to decrease even if the current density was kept constant [5, 10]. Both the critical void length and the metal height influence this phenomenon. The former is defined as the void length to create a measurable resistance increase in the test line, and it was found to coincide approximately with the via size. Due to a decrease in line height with every technology generation, the effective drift velocity under EM test conditions increases, leading to earlier failure. The scaling factor employed to reduce the interconnect dimensions of each new technology generation, $s$, is typically on the order of 0.7; i.e., the via sizes, metal heights and widths are reduced by approximately 30 %. The resulting EM lifetimes then scale with about $s^2$, which leads to a reduction of about 50 %. In addition to this geometric scaling behavior, transistor frequencies as well as front end and back end capacitances tend to increase the needed current densities for each generation. This effect results in even shorter lifetimes. As stated above, early failure mechanisms control the chip lifetime at operating conditions. Thus, it is necessary to evaluate the scaling behavior of early failure phenomena.

Figures 13(a) and (b) show TEM cross-sections of interconnects from the 90nm and 65nm technologies, respectively. The scaling factors for the line width and height are about 0.75 and 0.6, respectively. Lifetime distributions from down-flow Wheatstone Bridge tests for both generations are shown in Figure 14(a). The test structures used for 65nm samples were designed differently compared to the 90nm structures. Each sample contains 2448 interconnects compared to 920 in case of 90nm technology.

Thus, deconvolution of the data is necessary to adequately compare the EM performance. Figure 14(b) displays the deconvoluted lifetime distributions for both data sets. The ratio between the 65nm and 90nm lifetimes is ~0.5, which is in good agreement with the dimensional reductions, considering that the via sizes were scaled by about 0.77 in this particular case. It is interesting to note that the lifetimes of the 65nm interconnects show a tighter distribution compared to the C90 samples. It is possible that differences in tools

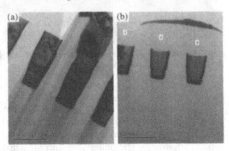

**FIGURE 13.** TEM images of (a) 90nm and (b) 65 nm technology interconnects.

and processes between the two technologies led to a cleaner interface between via and metal interconnect, thus reducing the risk for very early failures. Additional data points are needed though for further assessment.

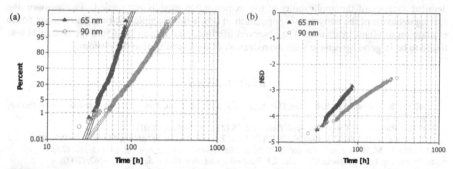

**FIGURE 14.** Lifetime distributions for down-flow Wheatstone Bridge devices for 90nm and 65nm technologies at 325 °C: (a) raw data, (b) deconvoluted data.

Interestingly, the 65nm samples yield an activation energy of 0.91 ± 0.04 eV between 275 °C and 325 °C. This value appears significantly higher compared to the activation energy of 0.83 ± 0.01 eV measured for the early failure mode in 90nm samples. The improved activation energy might also be due to newer tool sets and improved via processing. It needs to be noted though that for 65nm only a small temperature range was evaluated, namely 50°C compared to 175 °C for 90nm. Additional 65nm experiments are ongoing to extend the temperature range for better accuracy of the measurement. In general, both characteristics, tight distributions and high activation energy, are essential when extrapolating to use conditions, especially with dimensional scaling leading to further reduction in EM lifetimes.

## CONCLUSION

This study investigated in detail the early failure phenomenon in down-flow EM experiments conducted on Cu interconnects. The Wheatstone Bridge technique was employed to increase the examined interconnect size well past standard single link testing capabilities. The tested sample size in this study encompassed a total of 802,320 for the 90nm technology node and 306,000 interconnects using 65nm processing, increasing the total sample size past 1.1 million. This method was shown to adequately identify early failure modes in EM lifetime distributions. The activation energy corresponding to the 90nm early mode in down-flow interfaces was determined to be 0.83 ± 0.01 eV, significantly smaller than the late mode value of 0.91 ± 0.03 eV. While the up-flow early mode appeared to be the limiting interface at EM testing conditions, extrapolations to operating temperatures indicated that the down-flow early mode possibly controls the chip lifetime at operating conditions as a result of the decrease in activation energy. Thus, this study clearly shows the importance of a thorough investigation of the characteristics of EM induced early failure modes. In addition, short-length effects were examined for very small failure percentages. The Wheatstone Bridge structures were examined at two temperatures and linear extrapolation of inverse lifetime data result in the same critical product of $(jL)^* = 2900 \pm 400$

A/cm. In comparison, single link Contact/M1 data extrapolates to 3000 ± 500 A/cm. These results indicate that even at very small failure percentages, the critical product may remain steady at about 3000 A/cm for SiCOH-based dielectrics. However, more work and a more detailed analysis of the applicability of the extrapolation models are needed. Furthermore, the scaling behavior of the early failure population in down-flow samples was evaluated using 65nm in addition to 90nm structures. The observed lifetime reduction in the 65nm interconnects was found to be in good agreement with the reduction of the metallization dimensions.

## REFERENCES

1. C.-K. Hu, R. Rosenberg, H.S. Rathore, D.B. Nguyen, and B. Agarwala, *Proc. IEEE Int. Interconnect Technology Conf.*, 267-269, (1999)
2. C.S. Hau-Riege, C.V. Thompson, *Appl. Phys. Lett.* **78** (22), 3451-3453, (2001)
3. E.T. Ogawa, K.-D. Lee, V.A. Blaschke, P.S. Ho, *IEEE Transactions on Reliability*, **51** (4), 403-419, (2002)
4. M.A. Meyer, M. Herrmann, E. Langer, E. Zschech, *Microelectronics Engineering*, **64**, 375-382, (2002)
5. R. Rosenberg, D.C. Edelstein, C.-K. Hu, K.P. Rodbell, *Annu. Rev. Mater. Sci.*, **30**, 229-262, (2000)
6. M. Hauschildt, *Ph.D. Dissertation*, The University of Texas at Austin, 2005
7. M. Hauschildt, M. Gall, S. Thrasher, P. Justison, L. Michaelson, R. Hernandez, H. Kawasaki, and P.S. Ho, *AIP Conf. Proc. of Stress Induced Phenomena in Metallization: 8th Int. Workshop*, **817**, 164-174, (2006)
8. M. Hauschildt, M. Gall, S. Thrasher, P. Justison, L. Michaelson, R. Hernandez, H. Kawasaki, and P.S. Ho, *Appl. Phys. Let.*, **88**, 211907, (2006)
9. M. Hauschildt, M. Gall, S. Thrasher, P. Justison, R. Hernandez, H. Kawasaki, and P.S. Ho, *J. Appl. Phys*, **101**, 043523, (2007)
10. M. Gall, M. Hauschildt, P. Justison, K. Ramakrishna, R. Hernandez, M. Herrick, L. Michaelson, and H. Kawasaki, *Mater. Res. Soc. Symp. Proc.*, **914**, 305 (2006)
11. E.T. Ogawa, K.-D. Lee, H. Matsuhashi, K.-S. Ko, P.R. Justison, A.N. Ramamurthi, A.J. Bierwag, P.S. Ho, V.A. Blaschke, and R.H. Havemann, *Proc. of Int. Rel. Phys. Symp.*, 341-349, (2001)
12. J. Gill, T. Sullivan, S. Yankee, H. Barth, and A. v. Glasow, *Proc. Int. Rel. Phys. Symp.*, 298-304, (2002)
13. J.B. Lai, J.L. Yang, Y.P. Wang, S.H. Chang, R.L. Hwang, Y.S. Huang, and C.S. Hou, *Proc. Int. Sym. VLSI Tech., Sys. and Appl.*, 271-274, (2001)
14. B. Li, C. Christiansen, J. Gill, R. Filippi, T. Sullivan, and E. Yashchin, *Proc. Int. Rel. Phys. Symp.*, 115-122, (2006)
15. S.-C. Lee and A.S. Oates, *Proc. Int. Rel. Phys. Symp.*, 107-114, (2006)
16. M. Hauschildt, M. Gall, P. Justison, R. Hernandez, M. Herrick, *AIP Conf. Proc. of Stress Induced Phenomena in Metallization: 9th Int. Workshop*, **945**, 66-81, (2007)
17. M. Gall, *Ph.D. Dissertation*, The University of Texas at Austin, 1999
18. M. Gall, C. Capasso, D. Jawarani, R. Hernandez, H. Kawasaki, and P.S. Ho, *J. Appl. Phys*, 90 (2), 732-740, (2001)
19. K.-D. Lee and P.S. Ho, *IEEE Transactions on Dev. and Mat. Rel.*, **4** (2), 237-245, (2004)
20. H. Tsuchiya and S. Yokogawa, Microelectronics Reliability **46**, no.9-11, 1415-1420, (2006)
21. E.T. Ogawa, J.W. McPherson, J.A. Rosal, K.J. Dickerson, T.-C. Chiu, L.Y. Tsung, M.K. Jain, T.D. Bonifield, J.C. Ondrusek, W.R. McKee, *Proc. of Int. Rel. Phys. Symp.*, 312-321, (2002)
22. I.A. Blech, *J. Appl. Phys.* **47**, 1203-1208, (1976)
23. S. Thrasher, M. Gall, C. Capasso, P. Justison, R. Hernandez, T. Nguyen, H. Kawasaki, *AIP Conf. Proc. of Stress Induced Phenomena in Metallization: 7th Int. Workshop*, **741**, 165-172, (2004)
24. P.-C. Wang, R.G. Filippi, L.M. Gignac, *Proc. IEEE International Interconnect Technology Conference*, 253-265, (2001)
25. K.-D. Lee, *Ph.D. Dissertation*, The University of Texas at Austin, 2003
26. C. Christiansen, B. Li, J, Gill, *Proc. IEEE International Interconnect Technology Conference*, 114-116, (2008)
27. M. Shatzkes and J.R. Lloyd, *J. Appl. Phys*, **59**, 3890, (1986)
28. J.R. Lloyd, *J. Appl. Phys*, **69**, 7601, (1991)
29. J.R. Lloyd, *Microelectronics Reliability* **47**, no.9-11, 1468-1472, (2007)

Mater. Res. Soc. Symp. Proc. Vol. 1156 © 2009 Materials Research Society

# Effect of Dielectric Capping Layer on TDDB Lifetime of Copper Interconnects in SiOF

Jeff Gambino, Fen Chen, Steve Mongeon, Phil Pokrinchak, John He, Tom C. Lee, Mike Shinosky, Dave Mosher

IBM Microelectronics, 1000 River Street, Essex Junction, VT, 05452

## ABSTRACT

In this study, intralevel dielectric breakdown is studied for copper interconnects in an SiOF dielectric, capped with either SiN or SiCN. The leakage current is higher and the failure time of dielectric breakdown is shorter for an SiCN capping layer compared to an SiN capping layer. It is proposed that the dielectric breakdown of the integrated structure is limited by the interface between the capping layer and the SiOF dielectric. Lower lifetime for dielectric breakdown is observed for structures with an SiCN cap compared to structures with an SiN cap, due to higher leakage current in the SiCN. The higher leakage for an SiCN cap is consistent with results from planar metal-insulator-semiconductor capacitors.

## INTRODUCTION

Copper interconnects have gained wide acceptance in the microelectronics industry due to improved resistivity and reliability compared to Al interconnects [1]. More recently, low-k dielectrics such as SiCOH have been introduced, to further reduce the circuit delay [2]. Because of the lower modulus and lower fracture toughness of SiCOH compared to $SiO_2$, the Cu wires in the last dielectric layers are generally fabricated in $SiO_2$ or SiOF. The line-to-line capacitance of the upper wiring layers can be reduced by using an SiCN capping layer rather than an SiN capping layer, similar to what is done at the lower wiring layers [3].

The reliability of the dielectric surrounding the copper interconnects is assessed using time dependent dielectric breakdown (TDDB) measurements on comb-serpentine structures [4]. For damascene Cu integration, the main leakage path between Cu metal lines, and hence the location of the dielectric breakdown, is typically at the interface between the inter-level dielectric and the capping layer. Therefore, the processing of this interface is critical for achieving a robust TDDB reliability. For example, the TDDB lifetime is increased by using plasma treatments in $H_2$ or $NH_3$ prior to SiN cap layer deposition on $SiO_2$ [4,5]. Unfortunately, the $H_2$ plasma treatment also increases the resistance of Cu, converting the surface to a high resistivity CuSiCON layer [6]. Hence, it is desirable to use an alternate plasma clean, such as an Ar plasma, to minimize the increase in resistance [7].

There have been relatively few studies comparing the effect of SiN versus SiCN capping layers on TDDB. For capping layers on SiCOH dielectric, it has been observed that an SiN capping layer has either lower TDDB lifetime [8] or similar TDDB lifetime [9] as an SiCN capping layer. In contrast, for an SiOF dielectric, the TDDB lifetime was longer for an SiN cap than for an SiCN cap [10].

In this study, the dielectric breakdown is studied for different capping layers on an SiOF dielectric, comparing an Ar versus $H_2$ preclean, and SiN versus SiCN capping layers. It is

observed that TDDB lifetime is longer with an SiN cap compared to an SiCN cap, consistent
with previous results on SiOF dielectrics [10].

## EXPERIMENT

Samples were fabricated using a 0.13 μm CMOS process with via-first dual damascene Cu
in an SiOF dielectric (i.e. fluorinated TEOS) [11]. After M1 and M2 CMP, the wafers were
capped with either 50 nm SiN deposited using high density plasma chemical vapor deposition
(HDP CVD) [6,7] or 35 nm SiCN deposited by plasma enhanced chemical vapor deposition
(PECVD) [3,12] (Fig. 1). Different plasma treatments were used prior to barrier deposition to
remove Cu oxide from the surface of the copper wires; either an Ar plasma or an $H_2$ plasma prior
to SiN deposition [7], and an $NH_3$ plasma prior to SiCN deposition [12]. In addition, two
different SiCN processes were used with either low nitrogen or high nitrogen content (i.e. "low
N" or "high N"). The nitrogen content is in the range typically used for SiCN capping layers (15
to 25 at%) [3]. The "high N" SiCN capping layers have 10% more nitrogen than "low N" SiCN
capping layers.

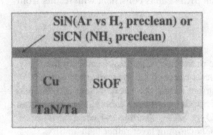

Fig. 1. Schematic of experiment. The
capping layer was either SiN (with Ar or $H_2$
preclean) or SiCN (with $NH_3$ preclean).

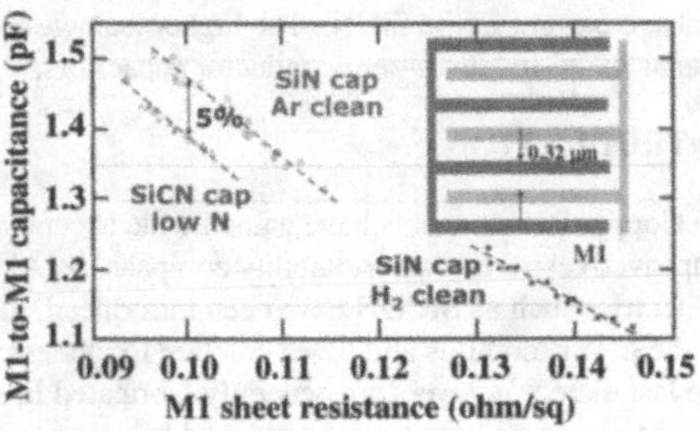

Fig. 2. M1 capacitance versus M1 resistance.

Metal sheet resistance and line-to-line capacitance were characterized using metal mazes
(minimum M1 pitch = 0.32 μm, minimum M2 pitch = 0.40 μm). The reliability of the dielectric
was measured for either M1 or M2 comb-comb structures. Constant voltage TDDB stresses were
measured at wafer level at 125°C with various voltages. The TDDB data were analyzed using
Weibull statistics [13, 14]. The structure of the patterned Cu wires was characterized by scanning
transmission electron microscopy (STEM).

In addition to the multi-level samples, simple metal-insulator-semiconductor (MIS) capacitors
were fabricated to investigate conduction and dielectric breakdown in SiN compared to SiCN.

## RESULTS

The M1 wire resistance is considerably higher with an $H_2$ plasma clean, compared to
either an Ar or an $NH_3$ plasma clean (Fig. 2), consistent with previous studies [6,7]. The M1
line-to-line capacitance is about 5% lower for the 35 nm SiCN capping layer compared to the 50
nm SiN (Ar clean) capping layer (Fig. 2).

The leakage current for M1 comb-comb structures shows a strong dependence on the capping layer (Fig. 3), indicating that the measured leakage is controlled by capping layer and/or the cap-SiOF interface, rather than by the bulk SiOF. For an SiN capping layer, the leakage is slightly higher for an Ar preclean compared to an $H_2$ preclean (Fig. 3a). For an SiCN capping layer, the leakage is much higher with the "low N" process compared to the "high N" process (Fig. 3b). Hence, the leakage for SiCN capping layers is either much higher than that for SiN (for SiCN with low N) or is only slightly higher than that for SiN (for SiCN with high N).

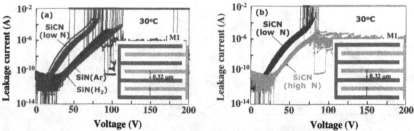

Fig. 3. Current-voltage characteristics for comb-comb structures with different capping layers; (a) SiCN (lowN), SiN with Ar preclean, SiN with $H_2$ preclean and (b) SiCN with low or high nitrogen content.

The breakdown voltage for M1 structures also shows a strong dependence on the capping layer (Fig. 3). For an SiN cap, the breakdown voltage shows a strong dependence on the preclean, ranging from 65 to 90 V for samples with the Ar preclean compared to 95 to 110V for samples with the $H_2$ preclean (Fig. 3a). With the SiCN capping layer, the dielectric breakdown depends on the nitrogen content, with a breakdown voltage between 65 and 80V for "low N", and a breakdown voltage of 80 to 90V for "high N" (Fig. 3b).

The TDDB lifetime is higher by about 100X for the SiN cap (Ar preclean) compared to the SiCN cap (low N) (Fig. 4), consistent with the leakage current and breakdown voltage data. The TDDB lifetime is even higher for the SiN cap with $H_2$ preclean.

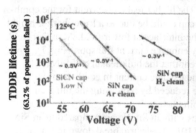

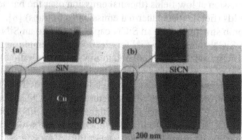

Fig. 4. TDDB data for M1 comb-comb structures with SiN or SiCN capping layers.

Fig. 5. STEM cross-sections of M2 wires with (a) SiN cap with $H_2$ preclean and (b) SiCN cap with low nitrogen content.

STEM images of the M2 structures show that the SiOF is recessed with respect to the top of the Cu wire for the SiN cap (H₂ preclean) whereas the SiOF is at the same level as the top of the Cu wire for the SiCN cap (Fig. 5).

The I-V characteristics of the MIS capacitors are dramatically different depending on the insulator (Fig. 6). The leakage current at high voltages is considerably lower for capacitors with an SiN dielectric compared to those with an SiCN dielectric. For capacitors with SiCN dielectrics, the leakage is slightly reduced by increasing the nitrogen content (not shown).

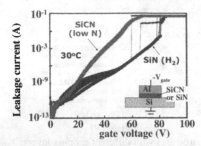
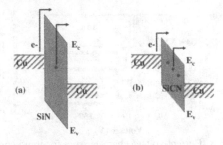

Fig. 6. Current-voltage data for MIM capacitors with 100 nm SiCN or SiN dielectric and Al gate electrode (negative gate bias).

Fig. 7. Schematic of band diagram at the metal-insulator interface for (a) SiN cap and (b) SiCN cap.

The I-V results of the MIS capacitors are consistent with the I-V results of the integrated comb-comb structures, suggesting that the leakage current for integrated structures is determined by the capping layer, rather than by the bulk SiOF dielectric.

## DISCUSSION

The conduction mechanism in comb-comb structures is usually dominated by Schottky emission at low fields (thermal emission over the barrier) and by Frenkel-Poole emission at high fields (field-assisted thermal emission from traps) [8]. The higher leakage current for the comb-comb structures with an SiCN cap compared to an SiN cap could be due to a lower barrier height of SiCN compared to SiN for electron injection along the interface at lower field. At higher field, more traps in SiCN compared to SiN could dramatically enhance current transport (Fig. 7). From the I-V characteristics of the MIS capacitors, it is apparent that the bulk leakage of SiN is significantly smaller than the bulk leakage of SiCN. The leakage current in comb-comb structures (Fig. 4) and MIS structures with SiCN can be reduced by increasing the nitrogen content, but the leakage is still higher than for SiN capping layers.

The lower TDDB lifetime for comb-comb structures with an SiCN cap compared to an SiN cap (Fig. 4) is consistent with the results of Aoki et al. [10]. Dielectric breakdown in Cu interconnect structures can be caused by Cu diffusion into the dielectric in the presence of a high electric field [13,14] or broken bonds in the dielectric due to electron injection [13,15]. Hence, lower TDDB lifetime for the SiCN cap could be due to one of the following: (1) faster Cu diffusion in SiCN than SiN (or faster Cu diffusion at the SiCN-SiOF interface compared to the

SiN-SiOF interface), (2) higher leakage current in SiCN than SiN, or (3) weaker bonds at the SiOF-SiCN interface than the SiOF-SiN interface. It is likely that the higher leakage current in structures with an SiCN cap compared to an SiN cap contributes to the shorter TDDB lifetime. The higher leakage current can lead to lower TDDB lifetime by damaging bonds in the insulators or by enhancing Cu ion diffusion into the dielectric (through electrochemical reactions). However, when looking at the data for the SiN cap samples, it is apparent that leakage current alone is not always an indicator of TDDB lifetime. For the structures with SiN caps and different precleans; the leakage current is almost the same for an Ar preclean versus an $H_2$ preclean, yet the TDDB lifetime at high voltages is much longer with the $H_2$ preclean. This result suggests that additional factors are important, such as Cu diffusion into the interface between the capping layer and the SiOF. Recent studies support the hypothesis that TDDB lifetime is determined by Cu diffusion into the dielectric and associated interfaces [13].

Based on this work and the previous studies, the TDDB lifetime is shorter for an SiCN cap (compared to an SiN cap) on SiOF [10], and is longer (or the same) for an SiCN cap (compared to an SiN cap) on SiCOH [8,9]. The different behavior between SiOF and SiCOH with respect to these capping layers suggests that the interface properties are more important than the bulk properties of the capping layer. For example, SiN capping layers may lead to more defects (dangling bonds) at the SiCOH-cap interface, whereas SiCN capping layers may lead to more defects (dangling bonds) at the SiOF-cap interface.

## CONCLUSION

In this study, intralevel dielectric leakage and breakdown are studied for Cu interconnects in an SiOF dielectric, capped with either SiN or SiCN. The leakage current is higher and the time dependent dielectric breakdown (TDDB) lifetime is shorter for an SiCN capping layer compared to an SiN capping layer. It is proposed that the lower TDDB lifetime for the SiCN cap is due to higher leakage current in the SiCN, consistent with results from MIS capacitors. The leakage current of the SiCN cap can be reduced by increasing the nitrogen content in the film, which is expected to improve the lifetime for dielectric breakdown.

## ACKNOWLEDGEMENT

The authors thank Carl Farris and the staff of the IBM Burlington manufacturing facility for assistance with processing the wafers. The authors also thank Son Nguyen for helpful discussions.

## REFERENCES

1. D. Edelstein, J. Heidenreich, R. Goldblatt, W. Cote, C. Uzoh, N. Lustig, P. Roper, T. McDevitt, W. Motsiff, A. Simon, J. Dukovic, R. Wachnik, H. Rathore, R. Schulz, L. Su, S. Luce, J. Slattery, IEDM Proc., 1997, p. 773.
2. D. Edelstein, C. Davis, L. Clevenger, M. Yoon, A. Cowley, T. Nogami, H. Rathore, B. Agarwala, S. Arai, A. Carbone, K. Chanda, S. Cohen, W. Cote, M. Cullinan, T. Dalton, S. Das, P. Davis, J. Demarest, D. Dunn, C. Dziobkowski, R. Filippi, J. Fitzsimmons, P. Flaitz, S. Gates, J. Gill, A. Grill, D. Hawken, K. Ida, D. Klaus, N. Klymko, M. Lane, S. Lane, J. Lee,

W. Landers, W.-K. Li, Y.-H. Lin, E. Liniger, X.-H. Liu, A. Madan, S. Malhotra, J. Martin, S. Molis, C. Muzzy, D. Nguyen, S. Nguyen, M. Ono, C. Parks, D. Questad, D. Restaino, A. Sakamoto, T. Shaw, Y. Shimooka, A. Simon, E. Simonyi, S. Tempest, T. Van Kleeck, S. Vogt, Y.-Y. Wang, W. Wille, J. Wright, C.-C. Yang, T. Ivers, IITC Proc., 2004, p. 214.

3. L.M. Matz, T. Tsui, E.R. Engbrecht, K. Taylor, G. Haase, S. Ajmera, R. Kuan, A. Griffin, R. Kraft, A. McKerrow, AMC 2005 Proc., MRS, 2006, p. 437.

4. J. Noguchi, N. Ohashi, T. Jimbo, H. Yamaguchi, K. Takeda, K. Hinode, IEEE Trans. Elec. Dev., **48**, 1340 (2001).

5. V.C. Ngwan, C. Zhu, A. Krishnamoorthy, Thin Sol. Films, **462-463**, 321 (2004).

6. A.K. Stamper, H. Baks, E. Cooney, L. Gignac, J. Gill, C.-K. Hu, T. Kane, E. Liniger, Y.-Y. Wang, J. Wynne, AMC Proc., 2005, MRS, 2006, p. 727.

7. J. Gambino, T.L. McDevitt, F.P. Anderson, J. Gill, S.A. Mongeon, J. Burnham, AMC 2006 Proc., MRS, 2007, p. 501.

8. A. Krishnamoorthy, N.Y. Huang, S.-Y. Chong, in Materials, Technology and Reliability for Advanced Interconnects and Low-k Dielectrics – 2003, MRS, vol. 766, 2003, p. 83.

9. T.Y. Tsui, R. Willecke, A.J. McKerrow, IITC Proc., 2003, p. 45.

10. H. Aoki, K. Torii, T. Oshima, J. Noguchi, U. Tanaka, H. Yamaguchi, T. Saito, N. Miura, T. Tamaru, N. Konishi, S. Uno, S. Morita, T. Fujii, K. Hinode, IEDM Proc., 2001, p. 76.

11. A.K. Stamper, C. Adams, X. Chen, C. Christiansen, E. Cooney, W. Cote, J. Gambino, J. Gill, S. Luce, T. McDevitt, B. Porth, T. Spooner, A. Winslow, R. Wistrom, AMC 2002 Proc., MRS, 2003, p. 485.

12. S. Sundararajan, M. Trivedi, US. Pat. # 6,444,568, 2002.

13. F. Chen, O. Bravo, K. Chanda, P. McLaughlin, T. Sullivan, J. Gill, J. Lloyd, R. Kontra, J. Aitken, IRPS Proc., 2006, p. 46.

14. F. Chen, B. Li, T. Lee, C. Christiansen, J. Gill, M. Angyal, M. Shinosky, C. Burke, W. Hasting, R. Austin, T. Sullivan, D. Badami, J. Aitken, IPFA Proc., 2006, p. 97.

15. J.R. Lloyd, E. Liniger, T.M. Shaw, J. Appl. Phys., **98**, 2005, p. 084109.

# Emerging Interconnect Technologies

Mater. Res. Soc. Symp. Proc. Vol. 1156 © 2009 Materials Research Society

# Direct Metal Nano-Patterning Using Embossed Solid Electrolyte

Anil Kumar, Keng Hsu, Kyle Jacobs, Placid Ferreira, and Nicholas Fang
Center for Nanoscale Chemical-Electrical-Mechanical Manufacturing Systems, University of
Illinois, Urbana-Champaign, IL 61801 U.S.A.

## ABSTRACT

In this letter we introduce a new approach to fabricating nano-scale metallic features by combining best merits of micro-forming, nanoimprint lithography, and electrochemical nano-imprinting. We study the mechanical properties of $Ag_2S$, a solid-state superionic conductor previously reported by Hsu et al. [1] for electrochemical nanoimprinting (S4 process), and explore its capability for embossing using a Si mold fabricated using electron-beam lithography. By circumventing the traditional route of stamp preparation using Focused Ion Beam (FIB), we greatly enhance the capability of electrochemical nanoimprinting. Using an embossed stamp, we demonstrate features <10 nm; and patterns on areas >30 $mm^2$. Application of an embossed $Ag_2S$ stamp is a significant step towards extending the S4 process for direct metal patterning of features beyond the capability of current processes.

## INTRODUCTION

Last few decades have seen a huge surge in nanotechnology which is expected to continue in future. This growth has been fueled by innovations in optoelectronics, nanoelectronics, nanooptics, nanoelectromechanical systems and various sensing applications including chemical and biological sensing. One of the basic building blocks of this growth is the ability to fabricate nanostructures with ever smaller features, good repeatability, and at lower cost and larger scales. Traditionally, photolithography has been the main process for fulfilling this demand. For features beyond the capability of photolithography, electron-beam lithography has been used [2, 3]. Recently, nano-imprint lithography has been very successful in scaling down the feature size while retaining the cost benefit [2-6]. It uses a Si or metal mold which is pressed (termed as microforming) into a photoresist followed by metal evaporation and lift-off.

These Si or metal mechanical parts, i.e., the molds are typically made by well-developed technologies such as micromachining, EDM, microforming, and etching (RIE) of e-beam patterns in case of imprint lithography. Attributed to its relatively low cost and ease in tooling and processing, microforming has recently received attention in further expanding its workable-material domain and reducing final feature dimensions. An intrinsic limitation of using this process to fabricate a large number of tiny metallic features over a large area lies in the force required to drive such an operation as well as the effect of global material flow on local features. Issues such as changes in mechanical properties of the features and undesired feature deformation can result. Therefore, the knowledge of mechanical properties of the surface to be microformed and its behavior during the process are important.

Here we report a new process to create nano-scale metallic features through the combination of the best merits of nanoimprint lithography, micro forming, and the solid-state electrochemical imprinting technique, S4, recently introduced by Hsu *et al.* [1, 6]. This process starts with the use of a micro-forming-like embossing process to engrave nano-scale features onto a solid electrolyte tool surface. This patterned solid electrolyte surface is then used to carry out the S4 process, creating a negative image on a metallic substrate. Figure 1 shows SEM images of a bowtie nanoantenna at different stages of the process. This process eliminates the costly Focused Ion Beam (FIB) milling used by Hsu *et al.* to create features on the electrolyte tool. It is also highly favorable for large-area patterning as well as mass-production of metallic substrates. The embossed solid-electrolyte tool surface can be easily trimmed off with microtome; the tool can then be re-used for embossing and patterning metallic substrates, reducing the feature dimension inconsistency across samples in large quantity. This is an attribute critical for applications in biomimetics and surface patterning for superhydrophobic structures. The ability to repeatedly use the mold, while being able to pattern large areas with smaller features offers significant advantages.

## EXPERIMENT

The Silicon molds for embossing were fabricated using electron-beam lithography. A 10 nm Cr layer was used as mask during reactive ion etching (RIE) of Si in $CHF_3$ atmosphere. The molds were etched to a depth of about 100 nm at an etching rate of 5 nm/min.

The $Ag_2S$ crystals synthesized with a furnace-reaction method described in [5] were machined to a conical tip. A diamond blade was then used to trim the very tip of the crystal off and form a small mesa of 300 μm in diameter with mirror-like surface finish (except for large area patterning shown later). One of these crystals was used for measuring the Young's modulus and hardness of $Ag_2S$ using a Hysitron Triboscope mounted on a Digital Instruments multimode system. A standard Berkovich diamond tip with a half angle of 65.35° and tip radius of 100-200 nm was used to indent a depth of 50-100 nm with an applied force of 100 to 200 μN (Figure 2).

Embossing was carried out using a custom design setup that allows weight of up to 500 lbs with precision of 0.1 lbs. After embossing, 40 nm thick Ag films with an adhesion layer of 3 nm Cr were used for the S4 process. The films were evaporated at a chamber pressure of $8 \times 10^{-7}$ Torr and a stable deposition rate of 1 Å/s. Results from different steps are shown in Figure 1.

Figure 1. SEM images from the different steps of the process. (a) Si mold fabricated using electron-beam lithography is used to emboss into the $Ag_2S$ stamp as shown in (b), which transfers the original pattern to a silver film as in (c).

## RESULTS AND DISCUSSION

One of the reasons, silver sulfide ($Ag_2S$) was chosen was because of its favorable mechanical properties for micro forming. Unlike most ionic conductors, $Ag_2S$ is ductile and has relatively low yield strength of 80 MPa. To explore the embossing behavior of $Ag_2S$ crystal mesa, we use nano-indentation, a common method to measure mechanical properties of small volumes. Figure 2 shows results from single- (grey/magenta diamonds) and multi-step (squares and circles) loadings on to a microtomed $Ag_2S$ surface under different loads. A reduced Young's modulus of 30 GPa and hardness of 0.6 GPa are observed. These values are comparable to corresponding values for metals like Sn, and Au. The elastic modulus is about four orders of magnitude higher than PDMS [8] (and about 10 times higher than PMMA), and the hardness is about two orders higher than PDMS; suggesting that some of the common issues related to stamp collapse [8] are highly improbable for $Ag_2S$. This comparison also gives an idea of the stress required for embossing $Ag_2S$: the modulus is sufficiently high to form patterns with well-defined features, but relatively low enough to not require serious investment in embossing equipment. To get a perspective, a 300 μm mesa required <0.1 lbs for embossing an array of bowtie structures with triangle edges of 400 nm and particle spacing of 3μm.

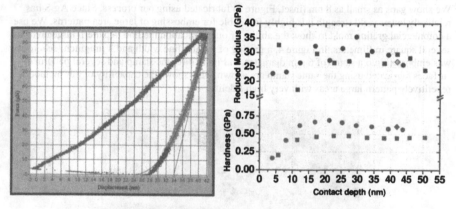

Figure 2. A single step loading-unloading curve during nano-indentation (left). The graph (right) shows the modulus and hardness obtained from three measurements. The grey and magenta diamonds are from the curve shown on the left, while the circles and squares are from two multi-step indentations at different locations. The crystal surface was microtomed using a diamond blade prior to nano-indentation.

In the loading curve, after an initial elastic region of approximately 10-12 nm, the slope changes as the material starts to yield. Unlike in the case of polymers [9], the slope in the second stage of plastic flow increases suggesting work hardening of the material as the tip further penetrates into the surface. An elastic recovery of ~15 nm can be observed from the unloading curve. This recovery is an important parameter for designing the molds, e.g. for etching 40 nm films the mold needs to be at least 55 nm high in order to avoid the bottom of features on stamp

143

contacting the film and etching it. Additionally, the surface quality of the crystal mesa at each step is very important because the embossed surface is further used to make the final metal features. The material flow after embossing can result into distortions in feature shape and bulging around edges. However, as the deeper parts of the mold come in contact with the stamp, bulging around features is no longer observed [10] and can be confirmed by AFM (not shown here) and the SEM image in Figure 1 (b).

The method of designing patterns on a stamp using FIB has several limitations that can be addressed using embossing. Apart from the requirement to pattern every new stamp, patterning large areas with FIB is very time consuming and expensive. Additionally, FIB has limitation in terms of smallest feature and largest area that can be patterned. Therefore, a Si mold using electron-beam lithography not only overcomes these limitations the mold can be post processed to get desired features. Two such examples of feature modifications are: (1) line width reduction using the lateral cutting during etching (RIE); and (2) line width increment using atomic layer deposition (ALD). Si molds also show good life span; our experiments so far have not shown any mold damage for several weeks and a few hundred embossing attempts.

Figure 3 shows a bowtie antenna with gap of 15 nm that was fabricated with high repeatability. Field enhancement in well designed narrow gaps of nanostructures like this are critical for biological and chemical sensing, and are currently an important area of research [11]. We show gaps as small as 8 nm (inset, Figure 3) fabricated using our process. Since $Ag_2S$ has relatively lower yield strength it is highly favorable for embossing of large area patterns. We use a commercial grating mold to show the ability of our fabrication method to pattern large areas of several square millimeters. In Figure 4, a grating of 1.5 μm line width and 3 μm pitch (see inset), was embossed into a stamp of 6 mm diameter. In Figure 4 (c) the stamp and several patterned surfaces fabricated using the same stamp are shown. This shows the capability of our method to repetitively pattern large areas with very high fidelity.

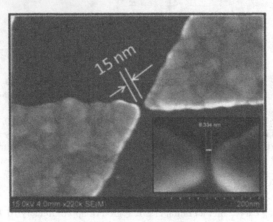

Figure 3. Silver bowtie nanoantennas with gaps of 15 nm were fabricated with very good repeatability using an embossed stamp. Nanoantenna gaps as small as 8 nm were fabricated (inset), which were previously not possible without embossing.

Figure 4. SEM images to demonstrate large area fabrication capability of the process. A commercial square grating with an edge of 6 mm (a), having a linear pattern of 1.5 μm line width and 3 μm pitch (inset (a)) was successfully transferred to a silver film (shown in (b) and (c)); the $Ag_2S$ stamp and several etched patterns can be seen in (c). The scale in (a) is 1 mm.

## CONCLUSIONS

We explored the capability of silver sulfide ($Ag_2S$) as a material conducive for embossing by studying its mechanical properties using nano-indentation. We demonstrated that several of the previously observed limitations of an FIB patterned $Ag_2S$ stamp can be successfully circumvented by embossing it using a Si mold. Embossing offers several unique advantages for direct metal nano-imprinting in terms of cost, time, repeatability, minimum feature size, and maximum area that can be patterned. Finally, using an embossed stamp, bowtie antennas with gaps as small as 8 nm were successfully fabricated; and patterning over area >30 sq. mm was achieved.

## ACKNOWLEDGMENTS

This work was carried out in part in the Frederick Seitz Materials Research Laboratory Central Facilities, University of Illinois, which are partially supported by the U.S. Department of Energy under grants DE-FG02-07ER46453 and DE-FG02-07ER46471

145

## REFERENCES

1. K. H. Hsu, P. L. Schultz, P. M. Ferreira, and N. X. Fang, *Nano Lett.* **7** (2), 446-451 (2007).
2. S. Y. Chou, P. R. Krauss, W. Zhang, L. Gou, and L. Zhuang, *J. Vac. Sci. Technol. B* **15**, 2897-2904 (1997).
3. I. Brodie and J. J. Muray, *The Physics of Micro/Nano Fabrication*, Springer-Verlag: New York, 1993.
4. Y. Wang, X. Liang, Y. Liang, and S. Y. Chou, *Nano Lett.* **8** (7), 1986-1990 (2008).
5. S. Y. Chou, and Q. Xia, *Nat. Nanotech.* **3**, pp. 295-308, 2008.
6. Molecular Imprints, Inc. www.molecularimprints.com; Nano Opt, www.nanoopto.com
7. P. L. Schultz, K. H. Hsu, N. X. Fang, and P. M. Ferreira, *J. Vac. Soc. Technol. B* **25**, 2419-2424 (2007).
8. Y. Y. Huang, W. Zhou, K. J. Hsia, E. Menard, J.-U. Park, J. A. Rogers, and A. G. Alleyne, *Lang.* **21**, 8058-8068 (2005).
9. H.-J. Butt, B. Cappella, and M. Kappl, *Surf. Sci. Rep.* **59**, 1-152 (2005).
10. M. Geiger, M. Kleiner, R. Eckstein, N. Tiesler, and U. Engel, *Anna. CIRP* **50**, 445-462 (2001).
11. S. Kim, J. Jin, Y.-J. Kim, I.-Y. Park, Y. Kim, and S.-W. Kim, *Nature* **453**, 757-760 (2008).

# Joint Session:
# Interconnect and Packaging

Mater. Res. Soc. Symp. Proc. Vol. 1156 © 2009 Materials Research Society          1156-D08-02-F06-02

# Low Temperature Direct Cu-Cu Immersion Bonding for 3D Integration

Rahul Agarwal[1] and Wouter Ruythooren[1]
[1]IMEC, 75 Kapeldreef,
Leuven, 3001, Belgium

## ABSTRACT

High yielding and high strength Cu-Cu thermo-compression bonds have been obtained at temperatures as low as 175°C. Plated Cu bumps are used for bonding, without any surface planarization step or plasma treatment, and bonding is performed at atmospheric condition. In this work the 25μm diameter bumps are used at a bump pitch of 100μm and 40μm. Low temperature bonding is achieved by using immersion bonding in citric acid. Citric acid provides in-situ cleaning of the Cu surface during the bonding process. The daisy chain electrical bonding yield ranges from 84%-100% depending on the bonding temperature and pressure.

## INTRODUCTION

Die-to-die stacking is a key enabler in 3-D integration with high density and high speed interconnections. At IMEC direct metal-to-metal bonding for die stacking is investigated as an alternative to solder bonding due to its advantages, such as low processing cost due to fewer processing steps and predictable reliability because of single metal joints. Unlike solder bonding [1,2], in direct metal-to-metal bonding there is no solder reflow which makes this technology very useful for tighter pitch bump formation. In this paper we present the results of the immersion thermo-compression bonding for direct (as plated) Cu-Cu interconnects. High yield and high strength direct Cu-Cu thermo-compression bonds are obtained at temperatures as low as 175°C and results from high density micro-bumps are presented.

Cu-Cu thermo-compression bonding requires higher temperature and pressure to make electrical connections as compared to solder bonding (for eg. Cu-Sn). In literature several processes have been discussed earlier where Cu-Cu metallic bonding is performed at lower temperature and/or lower pressure by conditioning the plated Cu bumps. Since there is no reflow of metal the surface roughness plays an important role and hence most of the low temperature Cu-Cu bonding results presented in literature have relied on surface planarization steps like CMP or diamond bit cutting to obtain a surface roughness of a few nanometers [3-5]. In the processes developed by Dr. Suga and his group the smooth metallic surface is first activated in plasma and bonding is performed at room temperature but at ultra high vacuum [6]. The disadvantage in all these processes is that an extra planarization step is required to make the plated surface smooth.

In the immersion bonding method presented here citric acid is present between the samples being bonded, providing in-situ cleaning of the Cu surface during the bonding. As plated Cu bumps with average roughness of more than 270nm are successfully bonded at temperatures as low as 175°C. Bonding is performed on two different test devices. First test device with 40μm pitch peripheral array (480 interconnections distributed over 2 daisy chains) have 100% yielding devices at temperatures as low as 175°C and 10g/bump load. The second test device with 100μm pitch area array (2018 interconnections distributed over 9 daisy chains) give a bump chain yield ranging from 84% to 100% depending on the bonding process conditions. For reference, samples which are bonded without citric acid clean prior to bonding did not show any cohesion while samples which are bonded after citric acid clean (but no in-situ cleaning) give only 44% electrically yielding daisy chains. Hence the results indicate an important strong beneficial impact of the immersion bonding method.

## SAMPLE FABRICATION

The test samples are designed so that after bonding the dies are connected to form a daisy chain allowing the assessment of electrical yield of the bonding process. Die to die bonding is performed on a SET FC6 flip chip bonder. Two different sets of 200mm wafers are processed to fabricate top die and landing dies. Figure 1 shows the schematic view of the fabrication process. Two different sets of 200mm wafers are processed to make landing die and top die. The processing starts with the deposition of a thin layer of dielectric on the wafers. Ti/Cu seed layer is then sputter coated on the wafer using a Nimbus 310 sputtering tool, as shown in figure 1(i). Next, first lithography is performed and Cu redistribution line (RDL) type tracks are plated on the wafers, as shown in figure 1(ii). On the landing die wire bond pads are also made out of these RDL tracks. After the plating the resist is stripped and second lithography is performed for bump plating, and the Cu bumps are plated using Stratus 100 plating tool. This is shown in figure 1(iii) and figure 1(iv). Then the photoresist is stripped in acetone/IPA and the seed layer is etched in a spray etcher as shown in figure 1(v).

Figure 2 shows the optical profilometer image of the Cu bump. As seen from the figure the average roughness on the bump is more than 270nm, which is much rougher than the one used by researchers in the past for low temperature Cu-Cu bonding.

The design of the wafer is such that we get a check board kind of pattern where every area array (AA) die is surrounded by peripheral array (PA) dies and vise-versa. The bump diameter is kept constant for the AA and PA patterns. The area array test devices have a total of 2018 interconnects at 100μm pitch. The bumps are distributed so that after bonding 9 different daisy chains are formed. After the bonding 4 pt. probe measurements are performed to determine the electrical yield of the device. This can be done on the whole device (2018 interconnects) or individual daisy chains can be measured to localize the failed section of the die. The PA dies have a bump pitch of 40μm, and there are only two interwoven daisy chains which can be measured individually. In total there are 480 bumps on a PA die.

## EXPERIMENTAL

Figure 3 shows the bonding profile used for this work. 120 seconds pressure ramp up time is used and the bonding is performed for 20 minutes at the peak temperature. Slow pressure ramp up time is used to allow enough time for citric acid to clean the oxides from the surface of the bumps. The time at peak temperature is not optimized and potentially there is a room to reduce it. Various different temperatures and pressures are used in this bonding experiment. Table 1 shows the design of experiments for this work. It is important to point out the device level and daisy chain level electrical yields. For example 7 area array dies are bonded at 250°C, 210MPa and 1 of them shows a 100% electrical yield for the daisy chains with in the die. This translates to a device yield of 14%. However, in the remaining six dies only one daisy chain failed out of 9 individual daisy chains. Hence with respect to the number of functional daisy chains, from a total of 63 bonded daisy chains of 225 bumps each, 56 daisy chains are electrically functional, thus giving a total daisy chain yield of 90%.

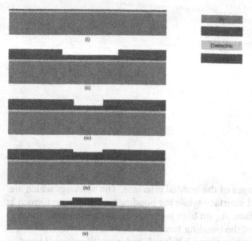

Figure 1: Schematic view of the fabrication process. (a) 200mm Si wafer with SiO2 insulation layer and Ti/Cu seed layer, (b) Patterned wafer with plated Cu tracks, (c) PVD Si3N4 is deposited on the wafer and pattered using dry etching using resist as the masking, (d) Third litho is performed to create bumps after sputtering the seed layers, (e) Finally the Cu bumps are plated and resist is removed followed by etching of the seed layer.

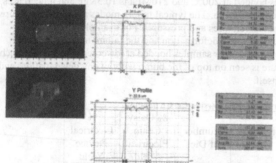

Figure 2: Optical profilometer scan of the as plated Cu bump.

In the area array dies the daisy chains failure was noticed at one particular corner. It might be due to the contamination of the bonding head or due to the bump non uniformity resulting from electroplating. A new tool was used for the peripheral array dies with 100% yield. On the AA dies it is also noticed that the samples which are bonded at lower pressure have less yield and they also shear at lower forces.

151

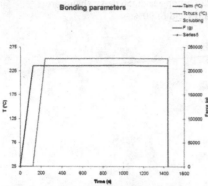

**Figure 3: Bonding profile used in this work.**

Figure 4 shows the SEM cross-section images of the bonded interface. The bondings which are performed at 250°C shows very good bond interface while the bondings which are performed at 200°C and 175°C shows voids at the interface. Again both temperature and pressure play very important roles in making a good contact of the bonding faces.

Shear tests are performed on number of samples using XYZtec Condor multifunctional bond tester to check for the bond integrity and observe the failure mode. A 100N force sensor is used in these measurements with a shear height of 20μm and shear speed of 10μm/sec. The shear test results are consistent with the trends observed in the electrical yield measurements. For example the average shear strength of samples bonded at 200°C and 210MPa is 10.5KgF and it decreases to 7.4KgF when bonding is performed at 100MPa. The typical failure modes are shown in figure 5 for bondings performed at different temperatures and pressures. For examples the samples which are bonded at 250°C shows severe plastic deformation and the failure mode is fracture through the bonded Cu-Cu interface; however for samples bonded at lower temperatures no such plastic deformation is seen. The failure is seen on top of the bump without any delamination, and very less deformation on the bump itself.

**Table 1: Electrical yield and shear tests results for immersion Cu-Cu bonding.**

| Number of dies bonded | Type of Dies | Temperature (°C) | Pressure (MPa) | Number of Dies | % Daisy Chain Electrical Yield | % Electrical Device Yield |
|---|---|---|---|---|---|---|
| 5 | AA | 250 | 210 | 5 | 100 | 100 |
| 10 | AA | 200 | 210 | 10 | 89 | 0 |
| 10 | AA | 200 | 100 | 10 | 84 | 0 |
| 7 | AA | 175 | 210 | 7 | 90 | 14 |
| 5 | PA | 250 | 210 | 5 | 100 | 100 |
| 5 | PA | 250 | 100 | 5 | 100 | 100 |
| 5 | PA | 200 | 210 | 5 | 100 | 100 |
| 5 | PA | 200 | 100 | 5 | 100 | 100 |
| 5 | PA | 175 | 210 | 5 | 100 | 100 |

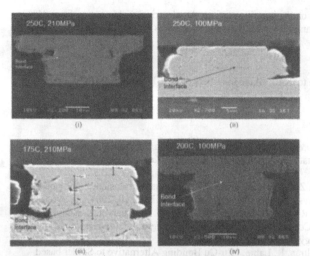

**Figure 4: Cross-section SEM image of the direct Cu-Cu bonding at various temperatures and pressures.**

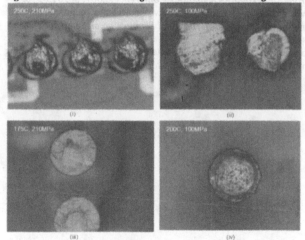

**Figure 5: Optical images of the sheared samples showing the typical mode of failure at different bonding temperature and pressures.**

## CONCLUSIONS

Direct Cu-Cu metal bonding is presented here. The bonding is performed on 'as plated' Cu bumps, with average roughness of more than 250nm, without any prior surface treatment. The bonding is performed in the presence of a cleaning agent, citric acid in this case, to clean and prevent the formation of oxide on the bonding surface. An electrical yield ranging from 84%-100% is recorded on various samples with bonding temperatures ranging from 175°C to 250°C.

## ACKNOWLEDGMENTS

The authors would like to thank all the members of the 3D team in IMEC for their support during this work.

## REFERENCES

1. W. Zhang, W. Ruythooren, "Study of the Au/In reaction for Transient liquid-phase bonding and 3D chip stacking," *Journal of Electronic Material*, vol. 37, no. 8, 2008, pp. 1095-1101.
2. N. S. Bosco, and W. F. Zok, "Critical interlayer thickness for transient liquid phase bonding in the Cu-Sn system," *Acta Materialia*, vol. 52, issue 10, 2004, pp. 2965-2972.
3. P. Gueguen, C. Di, M. Rivoire, D. Scevola, M. Zussy, A. M. Charvet, L. Bally, D. Laforn, L. Clavelier," Copper direct bonding for 3D integration," *International Interconnect Technology Conference, 2008,* pp. 61-63.
4. W. Ruythooren, A. Beltran, R. Labie, "Cu-Cu Bonding Alternative to Solder based Micro-Bumping," *Electronics Packaging Technology Conference*, 2007, pp. 315-318.
5. K. Arai, A. Kawai, M. Mizukoshi, H. Uchimura, "A new planarization technique by high precision diamond cutting for packaging", *International Symposium on Semiconductor Manufacturing*, 2004, pp. 534-537.
6. T. H. Kim, M. M. R. Howlarer, T. Itoh, T. Suga, "Low temperature direct Cu-Cu bonding with low energy ion activation method," *Electronic Materials and Packaging*, 2001, pp. 193-195.

Mater. Res. Soc. Symp. Proc. Vol. 1156 © 2009 Materials Research Society          1156-D08-04-F06-04

# Failure Analysis and Process Improvement for Through Silicon Via Interconnects

Bivragh Majeed, Marc Van Cauwenberghe, Deniz S. Tezcan, Philippe Soussan
IMEC
Kapeldreef 75, B-3001, Leuven, Belgium

## ABSTRACT

Through silicon vias (TSV) is one of the key enabling technologies for 3D wafer level packaging (WLP). This paper investigates the failure causes of TSVs in a "via last" approach and presents process improvement for implementing the TSV. There are many parameter including silicon etch uniformity, dielectric etching at the bottom of the TSV, non-uniform plating and resist residue inside the via that can reduce the yield of the process. We report that one of the main factors contributing to the yield loss is silicon dry etching effects including non-uniformity and notching. A new via shape that is a combination of sloped and straight etching sequence is developed in order to improve silicon dry etch effects. An improved and characterized, notch free uniform silicon etching across the wafer process based on three step etching is presented. An integration flow implementing the optimized parameters with showing electrical interconnection is given in the paper.

## INTRODUCTION

IMEC is investigating two generic approaches for 3D wafer integration that are categorized into via first, 3D SiC, and via last, 3D WLP, strategies [1]. 3D technologies have become one of the main drivers for continuous system miniaturization and driving "More than Moore" applications. [2]. "More than Moore" concept suggests new integration techniques that combine multiple die/wafer with multiple functions into a single module rather than trying to combine vastly different types of circuits on a single die. The main advantage of this approach is to bring together different IC's and build highly integrated modules with optimized performance, size and cost. TSV are one of the key technologies for realizing "More than Moore" concept for different 3D integration routes. At IMEC, 3D SiC approach aims to create very high density vertical interconnects to allow stacking of die. This approach also referred to as via first, as via are formed during the CMOS processing prior to BEOL. It involves formation of silicon via, dielectric deposition, copper filling, standard BEOL processing followed by bonding the device wafer to a carrier, thinning down to 20 microns and a CMP step to expose the copper plug. These plugs are then used for make interconnection between different dice. IMEC has demonstrated yielding 10K TSV chains with an average via pitch of 10 microns for a via diameter of 5 microns [3]. 3D WLP approaches at IMEC can be divided into two categories depending on the scalability of the process. Both are based on "via last" methodology where the TSV are processed after CMOS and even after wafer thinning process. Hence it is a generic approach and compatible with standard IC wafers allowing direct stacking of wafers or die. We have previously demonstrated yielding via chains of different size ranging from 50-150 microns based on sloped TSV with CVD deposited Parylene and conformal copper plating [4, 5]. In the second approach, a scalable TSV process based on a donut etch with spin on polymer filling and bottom up copper via filling has been demonstrated [6]. In this paper we investigated the failure causes for the sloped TSV and suggest a new via shape to improve the yield of the process.

## PROCESS FLOW

The process flow for the TSV approach is given in Figure 1. The process can be divided into three main categories. The first part, referred as front side processing, oxide, TaN and aluminum layers are deposited on a 200mm wafer (a), followed by resist patterning and etching of aluminum and TaN layers (b). This stack mimics the standard landing pads in MCM-D platform at IMEC. In the second phase, referred as bonding and thinning, the processed wafers are bonded onto a glass carrier with a temporary glue layer .This is followed by thinning process to achieve the final silicon thickness of 100 microns (c, d). The two above processes will not be discussed further in this paper as they have been standardized and reported elsewhere [4, 5, and 7]. The last block is the TSV processing and this part will be the main focus in this paper. The process starts with the etching of TSV in the bonded wafer landing on oxide layer at the bottom of TSV (e). The oxide is then etched followed by dielectric deposition in the TSV. The improvement in dielectric deposition is still under investigation and will not be discussed here and the current flow is without dielectric layer. A seed is then sputtered and a negative resist is spun on the wafer (f, g). A conformal copper is plated in the TSV to obtain the electrical interconnections. The plating is done in a two step process (h). First, a low-current density is applied to have a gradual deposition, which is done to enhance the sputtered Cu layer on the sidewalls. In the second step of the process, a higher current density is used to do the bulk of the Cu plating. After the plating the resist is removed and the seed layer is etched to complete the TSV processing (i).

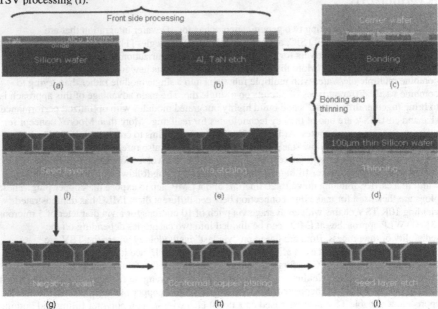

**Figure 1:** Integration process flow for TSV development (a, b) front side processing, (c, d) bonding and thinning, (e-i) TSV processing.

## FAILURE ANALYSIS

After the first integration process flow, the wafers were electrically measured and the yield of the process was found to be very low. There are many parameter including silicon etch uniformity, oxide etching at the bottom of the TSV, non uniform plating and resist residue inside the TSV that can reduce the yield of the process. Different failure analysis techniques were investigated to determine the cause of failure. FIB cross-sectional analysis provided the best results. The results are shown in Figure 2. Figure 2(a) shows a general overview of the bottom half of the TSV and shows notching at the bottom of the TSV. It also shows that copper is plated on the side wall uniformly and even plated at the bottom. The notching cause a discontinuity in the seed layer resulting in no plating at the bottom interface and no connection is obtained between the top and bottom contacts. This is emphasized in the zoom image Figure 2(b). It also shows that front side stack of oxide, TaN and Al layers and oxide is etched at the bottom of the TSV. With conventional Bosch process, notching at the interface between oxide and silicon has been observed. In our process, the wafers are bonded and the silicon is ground; this introduces an inherent thickness variation across the wafer. At the same time, the etching itself can be non-uniform across the wafer. To overcome these problems, there is a need for over-etching of silicon. The combine effects of the thickness variation, the etching non-uniformity and the over etch exaggerate notching at the oxide silicon interface.

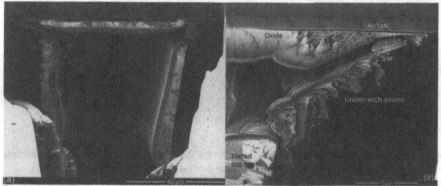

**Figure 2**: FIB cross-sectional images of (a) general over view of TSV (b)

From the above discussion it was concluded that, in our case, the notching was the main criterion for the failure of TSV and further investigations were needed to optimize the process. The notching occurs due to positive potential build up at the bottom of the TSV over a number of rf cycles during the deep reactive ion etching. This build up causes the trajectory of ions to be distorted; and the side walls at the bottom of the TSV are attacked further, causing the notching [8]. There have been some reported studies that looked at the charging effect during plasma etching. Arnold, demonstrated that localized charging could cause significant potentials that could tilt the ion directionality on a insulating surface in RIE[9]. While Kim described a method to prevent silicon from notching by introducing self-aligned metal interlayer to prevent charge buildup [10]. However, this process requires additional processing on the front side of the wafers. In this paper we focus on optimizing the etching step by changing the etching conditions and chemistry to minimize the notching.

## EXPERIMENTAL DETAILS

The main focus of the current experiment is to optimize the etching parameters. We report on a new TSV shape that is a combination of sloped and straight etching sequence to form "Chamfered TSV". In etching, first a sloped etching is optimized with changing different etching parameters. The etching recipe consists of SF6, C4F8 and O2 gasses and flow rate were adjusted to provide a side wall with a slope angle of 60°. Sloped TSV facilitates in subsequent dielectric deposition and sputtering processes. Secondly, a vertical etching process based on Bosch technique is optimized. The initial process was based on a standard Bosch including etch and passivation step, while later on a soft landing step with longer passivation steps were investigated. The resulting chamfered TSV on the blanket silicon is given in Figure 3.

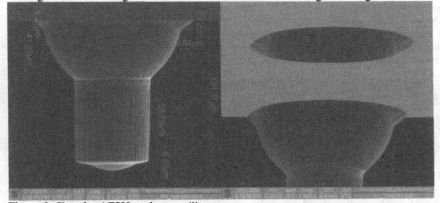

**Figure 3**: Chamfered TSV on dummy silicon

After initial investigation on dummy silicon wafer, the thinned bonded wafers were used. Silicon test wafers with front side processing and thinning process as given in Figure 1(a-d) were prepared. Table 1 shows the different etching recipes that were investigated in the work. Different parameters that influence on the TSV etching including influence of grinding marks, wafer thickness variation, etching rate and etching profile across the wafer were investigated.

Table 1: Different test conditions for etch optimization experiments

| Wafer ID | D06 | D07 | D24 | D02 | D04 |
|---|---|---|---|---|---|
| Step 1 | 15 min. tapered etch | 15 min. tapered etch | 15 min. tapered etch | 15 min. tapered etch | 15 min. tapered etch |
| Step 2 | 19 min. anisotropic etch | 20 sec. residue cleaning etch | 15 min. anisotropic etch | 30 min. soft landing etch | 13 min. Hi LF bias anisotropic etch |
| Step 3 | | 19 min. anisotropic etch | 5 min soft landing etch | | 10 min. soft landing etch |

## Results and Discussion

The test wafers were thinned down to 110 microns. The grinding is done in two steps: first a rough grinding to reduce bulk of the thickness and finally a fine grinding to obtain the final thickness. The grinding process leaves a footprint of grinding line across the wafer and top opening of the TSV is influenced by these grinding marks. The results showed that grinding marks cause the circular etching to be elongated along the marks as shown in Figure 3. To overcome this problem a dry recess etch can be preformed to reduce the grinding damage. The next parameter analyzed was thickness of the wafers. Thickness of the wafer was analyzed by two techniques and results were compared. In the first method, wafers were measured using an optical technique. In the second method, after etching the thin samples were mounted in epoxy and polished. Thickness of wafer was then measured using optical x-section. The results show that the two measurement techniques are very comparable as given in Table 2. The total thickness variation across the wafer is less than 2 micron. It also shows that grinding process is quite stable and repeatable.

**Table 2:** Wafer thickness measurement across different wafers

| Wafer ID | Technique | D06 | D07 | D24 | D02 | D04 |
|---|---|---|---|---|---|---|
| Average thickness ($\mu$m) | X-section | 107.7 | 109.9 | 108.0 | 108.4 | 109.8 |
| Standard deviation | | 0.6 | 0.5 | 0.5 | 0.7 | 0.6 |
| Average thickness ($\mu$m) | Optical | 110 | 109.4 | 109.7 | 108.4 | 110.3 |
| Total thickness variation | | 1.8 | 1.6 | 1.0 | 2.0 | 1.3 |

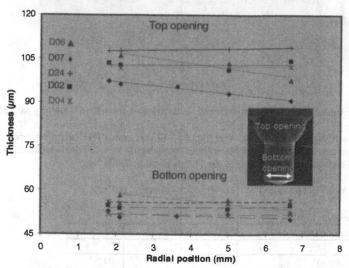

**Figure 4:** Top and bottom opening of TSV as a function of wafer radial position

The next parameters analyzed were top and bottom openings of TSV. Figure 4 shows the top and bottom openings as function of radial position on the wafer. For the two wafers (D06-D07) that use non-optimized parameters there is variation across the wafer for both the openings but for the optimized process it can be seen that variation across the wafers is considerably decreased (D24, D02, D04). Variation of top and bottom opening between different wafers is attributed to the variation of resist openings before the etching.

Based on the etch depth and etching time etch rates were calculated. It was found that isotropic etch rate in range of 2.5-3 microns/min, while etching rate for Bosch process is 3-3.5 microns/min and etch rate for soft landing step is 2.2 microns/min. The last parameter analyzed is the under/over etch during the process. The uniformity of etch rate and notching across the wafer were assessed and results are shown in Figure 5. A positive value indicates under-etch while a negative value indicate over etch and would result in notching. As can be seen for original process the etching rate is not uniform across the wafer. In the centre the TSV have landed on oxide, while at the edge there is still some silicon left. While in case of optimized process of using a soft landing step, the etch rate is quite uniform across the wafer. The samples were cross-sectioned and no notching was observed.

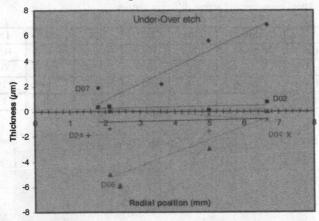

**Figure 5:** Under-over etch as a function of radial position for different etching parameters

Based on these results an integration lot was processed. The wafers were processed as given in process flow in Figure 1 but without the dielectric step and then electrically measured. The yield and electrical resistance of the TSV is given in Table 3.

**Table 3**: Yield and average TSV resistance for optimized process

| TSV dimension (µm) | 75 | 100 | 150 |
|---|---|---|---|
| Total # Daisy chains= 310* | 238 | 252 | 287 |
| Yield | 77 | 81 | 97 |
| Average (ohms) | 0.244 | 0.070 | 0.042 |
| * A daisy chain consist on average of 18 TSV | | | |

160

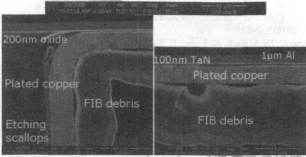

**Figure 6**: FIB cross-section of yielding TSV showing a notch free etching and very conformal copper plating

Figure 6 shows the FIB cross-sectional image of an electrically yielding TSV interconnect. The image shows that there is no notching at the bottom of TSV. The zoom image show that oxide is etched at the bottom of the TSV and copper is conformally plated in TSV.

## CONCLUSIONS

This paper reports the failure analysis of TSV and discuses the processes improvement to obtain higher yielding TSV. It was concluded that notching at the bottom of TSV was the main cause of failure in our process. We report on a new TSV shape that is a combination of sloped and straight etching sequence. Different parameters including influence of grinding marks, wafer thickness variation, etching rate and etching profile across the wafer were investigated. An improved and characterized process based in three step etching will was presented. The optimized etching sequence resulted in notch free uniform silicon etching and electrically yielding TSV across the 200mm wafer.

## REFERENCES

[1] E. Beyne and B. Swinnen, *Integrated Circuit Design and Technology, 2007, ICICDT '07*, (2007), 1-3.
[2] S. Mukherjee, R. M. Aarts, R. Roovers, F. Widdershoven and M. Ouwerkerk, "*Interconnect and Packaging Technologies for Realizing Miniaturized Smart Devices*", Philips Research, **5**,

AmIware Hardware Technology Drivers of Ambient Intelligence, Springer Netherlands, (2006)

[3] B Swinnen, W Ruythooren, P. De Moor, L. Bogaerts, L. Carbonell, K. De Munck K, B. Eyckens, S. Stoukatch, D. S. Tezcan, Z. Tokei, J. Vaes, J. Aelst and E. Beyne, *Proc. IEDM Conf.* San Fransisco, LA, (December, 2006), 1-4

[4] D.S. Tezcan, N. Pham, B. Majeed, P. De Moor, W. Ruythooren and K. Baert, *Proc. 57th Elec. Comp. Tech. Conf.*, Reno, NV, USA, (June 2007), 643-647

[5] B. Majeed, N. P. Pham, D. S. Tezcan, and E. Beyne, *Proc. 58th Elec. Comp. Tech. Conf.*, Orlando, FL, USA, (May 2008), 1556-1561,.

[6] D.S. Tezcan, F. Duval, O. Luhn, P. Soussan and B. Swinnen, *Int. Conf. Solid State Dev. Mat.* Japan, (Sep. 2008), 52-53

[7] R. C. Teixeira, K. De Munck, K. Baert, B. Swinnen, A. Knüttel and P. De Moor, *9th IEEE Electr Packag Technol Conf, Singapore* (Dec. 2007), 238-241.

[8] T. Kinoshita, M. Hane and J. P. McVittie, *J. Vac. Sci. Technol B,* 14 560-565 (1997).

[9] J. C. Arnold and H. H. Sawin, J. Appl. Phys. **70**, 5314-5317 (1991).

[10] C. H. Kim and Y. K. Kim, J. Micromech. Microeng. **15**, 358-361 (2005).

Mater. Res. Soc. Symp. Proc. Vol. 1156 © 2009 Materials Research Society    1156-D08-05-F06-05

# Effects of Thinned Multi-Stacked Wafer Thickness on Stress Distribution in the Wafer-on-a-Wafer (WOW) Structure

H. Kitada[1], N. Maeda[1], K. Fujimoto[2], K. Suzuki[2], T. Nakamura[3] and T. Ohba[1]

[1]University of Tokyo, 2-11-16 Yayoi, Bunkyo-ku, Tokyo 113-8656, Japan

[2]Dai Nippon Printing, 250-1 Wakashiba, Kashiwa-shi, Chiba 277-0871, Japan

[3]Fujitsu Laboratories Ltd., 10-1 Morinosato-Wakamiya, Atsugi, Kanagawa 243-0197, Japan

## ABSTRACT

In the trough silicon via (TSV) structure for 3-dimensional integration (3DI), large thermal-mechanical stress acts in the TSV caused by the mismatch in thermal expansion coefficient (CTE) of the TSV materials. In this study, the stress of multi-stacked thin silicon wafers composed of copper TSV and copper/low-k BEOL structure was analyzed by the finite element method (FEM), aiming to reduce the stress of TSV of 3D-IC. The results of sensitivity analysis using design of experiment (DOE) indicated that the thickness of the silicon and adhesive layer are the key factors for the structural integration of TSV design.

## INTRODUCTION

Since conventional downsizing based on an empirical Moore's law has reached the limitations of manufacturability, performance, and power consumption, an alternative integration technology such as 3DI is needed. There have been many reports on 3DI involving many wafer bonding techniques [1-4]. The TSV, which is larger than the BEOL interconnects, is used for the interconnects in 3DI, But the large CTE mismatch of the adhesive material used for wafer bonding, Si substrate, and copper influence the stacked TSV structure. Recently, a roadmap for

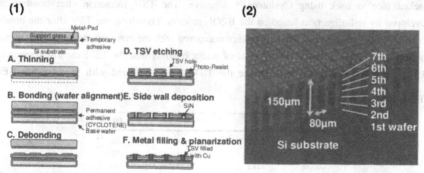

Fig. 1. (1) WOW process flow and (2) Seven-stacked thin wafers with TSV structure fabricated by the WOW process [6,7].

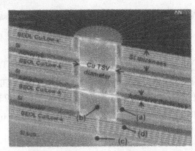

Fig. 2. FEM analysis model of 4-layer stacked Si with BEOL structure and copper TSV.

the TSV diameter and aspect ratio of 3DI was presented by the ITRS. However, the roadmap approach is not suitable for adhesive material in 3DI because there are a lot of purposes such as logic/memoly, MEMS, and hetero device stacking. It was reported based on FEM thermal stress analysis around the TSV [5]. However, the influence of TSV stress that an aspect ratio has not been discussed in 3D-IC structures that use the adhesive. In this report, the sensitivity to copper stress of TSV aspect ratio and adhesive layer such as higher CTE was estimated by DOE using FEM analysis in order to eliminate the stress in TSV structures.

## EXPERIMENTS

A novel through-silicon via (TSV) integration process formed after wafer bonding based on WOW has been developed [6,7]. The WOW process offers wafer-scale 3D manufacturability and high productivity of chip integration. Because the WOW uses a thinned Si wafer (< 20μm) stacking process, the TSV dry etching and copper plating process can be simplified. Figure 1 shows a prototype seven-layer stacked Si/TSV fabricated by the WOW process. The wafer was stacked face to back using Cyclotene™ adhesive. The TSV formation after bonding was developed by self-alignment based on the BEOL process. Therefore, the TSV after the bonding process can be processed at low temperature compared with the conventional metal-metal bump connection process. In the WOW process of wafer bonding and the TSV, the low-temperature (250°C) process was expected to reduce the TSV stress compared with the common BEOL

Table 1. Material properties.

| Material | E (GPa) | Poisson Ratio | CTE (ppm K⁻¹) |
|----------|---------|---------------|----------------|
| Silicon | 130.2 | 0.28 | 2.6 |
| Adhesive | 2.9 | 0.34 | 52 |
| Copper | 127.5 | 0.33 | 16.6 |
| Low-k | 8.0 | 0.20 | 16 |
| SiN | 140 | 0.25 | 2.8 |

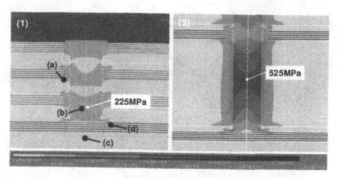

Fig. 3. Dependence of copper via height of maximum principal stress of via of different heights. Four-layer stacked wafer with 1) 20-μm thick wafer and 2) 100-μm thick wafer.

process (400°C).

By employing a model for the TSV using rotational symmetry as shown in Fig. 2, FEM analysis was used for the thermal stress analysis. The material properties are shown in Table 1. The Si substrate with a simple copper/low-k BEOL layer of 10-μm thickness was stacked to four layers, and the TSV structure was connected with a copper via. The residual stress was calculated at room temperature assuming a free stress at 250°C of copper and adhesive.

**RESULTS AND DISCUSSION**

Figure 3 shows the results of FEM analysis of the maximum principal stress for different silicon thicknesses. The stress inside the copper via with 100-μm thick Si was about twice that of the thinner Si (20-μm) with 30-μm constant diameter via. The copper stress at the high aspect

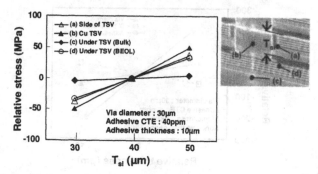

Fig. 4. Stress sensitivity analysis result of critical points from (a) to (d) around via. Si thickness has the largest effect on stress in copper via.

165

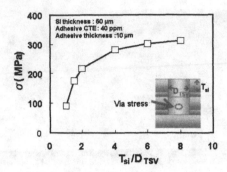

Fig. 5. Dependence of via stress on $T_{Si}/D_{TSV}$ by DOE analysis. Si thickness in the case of constant $T_{Si}$. $T_{Si}/D_{TSV}$ ratio has the largest effect on stress in copper via.

ratio via had exceeded the yield stress (286MPa) of copper.

The stress distribution shows that there are the following critical points: (a) BEOL region on via side, (b) inside via, (c) the bulk region under via and (d) BEOL under via. The influence of Si thickness ($T_{Si}$), TSV diameter ($D_{TSV}$) and adhesive thickness ($T_{adhesive}$) were analyzed by sensitivity analysis on the DOE method. Figure 4 shows results of stress sensitivity analysis for each factor at the critical points. A structural variation was given to a typical condition that there was an assumption of 40ppm CTE, 30-μm diameter, 40-μm Si thickness and 10-μm adhesive layer thickness. A tensile stress increase in copper TSV works with a smaller diameter and thinner silicone. Figure 5 shows the dependence of stress on the $T_{Si}/D_{TSV}$ ratio calculated using Chevyshev's orthogonal function obtained by DOE analysis. The tensile stress as the yield stress (286 MPa) hardly acts upon a TSV by increasing the $T_{Si}/D_{TSV}$ ratio. The thick adhesive material

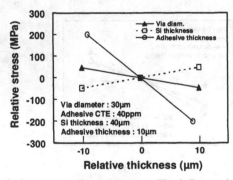

Fig. 6. Dependence of via stress on $T_{Si}$ and $T_{adhesive}$. The influence level of adhesive layer thickness was higher about four times the Si thickness

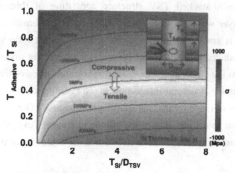

Fig. 7. Relation of the $T_{Si}/D_{TSV}$ and $T_{adhesive}/T_{Si}$ of the copper via stress. A thinner adhesive layer yields a lower stress TSV structure in addition to the effectiveness of having a lower aspect ratio via.

such as high CTE gave the compressive stress contrary to Silicon, and the influence level was about four times the Si thickness as shown in Fig. 6. It was clear that the $T_{adhesive}$ also had a large influence on the stress in TSV. Figure 7 shows the relation of $T_{Si}/D_{TSV}$ ratio and $T_{Adhesive}/T_{Si}$ ratio of the stress in the copper via. Both the diameter of TSV and silicone thickness was closely related to the stress in copper via, and the stress was smaller with a lower $T_{Si}/D_{TSV}$ ratio and thinner adhesive layer. This result shows that a thinner adhesive layer yields a lower stress TSV structure in addition to the effectiveness of having lower aspect ratio via.

These analytical results of features dependence suggest that the optimum adhesive thickness should be chosen according to the TSV aspect ratio to consider the yield stress of copper. Using the WOW process, multi-stacked 20-μm thinned Silicon wafers with 30-μm diameter and 80-μm pitch TSVs as shown in fig. 8. It was demonstrated that the WOW structure that has a low aspect via ($T_{Si}/D_{TSV}=0.67$) and thin adhesive ratio ($T_{adhesive}/T_{Si}=0.25$) was almost a free stress structure by the stress analysis.

## CONCLUSION

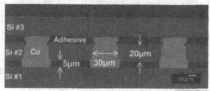

Fig. 8. Cross section SEM image of lower stress via structure. Stacked thin (20-μm) wafer and TSV (30-μm diameter) structure fabricated by the WOW process.

The stress of a multi-stacked TSV structure consisting of thinned silicon wafers with copper/low-k BEOL layer was analyzed using FEM and DOE sensitivity analysis. The results suggest that via stress is smaller with the case of a lower aspect ratio via structure. When the silicon thickness was reduced to 20 μm from 100 μm, the copper via stress decreased to about 1/2. And it was clarified that the stress also was influenced by the adhesive thickness. One of the key findings of this analysis for structural integration of TSV design is that stress is affected by not only the via aspect ratio but also by the adhesive layer thickness. The sensitivity of the adhesive thickness was high about four times that of the silicon thickness. An optimal design of a vertical structure of TSV that can control the stress will become important to form low stress of copper in a 3DI.

## ACKNOWLEDGMENTS

We would like to thank Nissan Chemical for its technical support. This work was carried out in the WOW Program.

## REFERENCES

[1] A. Jourdain, S. Stoukatch, P. De Moor, W. Ruythooren, S. Pargfrieder, B. Swinnen, and E. Beyne. *Proceeding of International Interconnect Technology Conference*, (2007) p 207.

[2] F. Liu, R. R. Yu, A. M. Young, J. P. Doyle, X. Wang, L. Shi, K.-N. Chen, X. Li, D. A. Dipaola, D. Brown, C. T. Ryan, J. A. Hagan, K. H. Wong, M. Lu, X. Gu, N. R. Klymko, E. D. Perfecto, A. G. Merryman, K. A. Kelly, S. Purushothaman, S. J. Koester, R. Wisnieff, and W. Haensch *Technical digest of* IEDM, (2008) p 599.

[3] Y. Kurita, S. Matsui, N. Takahashi, K. Soejima, M. Komuro, M. Itou, C. Kakegawa, M. Kawano, Y. Egawa, Y. Saeki, H. Kikuchi, O. Kato, A. Yanagisawa, T. Mitsuhashi, M. Ishino, K. Shibata, S. Uchiyama, J. Yamada, and H. Ikeda. *Proceeding of* ECTC, (2007) p 821.

[4] P. Leduc, F. de Crécy, M. Fayolle, B. Charlet, T. Enot, M. Zussy, B. Jones, J.-C. Barbé, N. Kernevez, N. Sillon, S. Maitrejean, D. Louis, G. Passemard. *Proceeding of International Interconnect Technology Conference*, (2007) p 210.

[5] C. Okoro, Y. Yang, B. Vandevelde, B. Swinnen, D. Vandepitte, B. Verlinden, I. De Wolf,. *Proceeding of International Interconnect Technology Conference*, (2008) p 16.

[6] N. Maeda, H. Kitada, K. Fujimoto, K. Suzuki, T. Nakamura, and T. Ohba, *Proc. of Advanced Metallization Conference*, (2008) p 91.

[7] T. Ohba, N. Maeda, H. Kitada, K. Fujimoto, K. Suzuki, T. Nakamura, A. Kawai, K. Arai. *Proceeding of Material for Advanced Metallization* (2009) p 127.

Mater. Res. Soc. Symp. Proc. Vol. 1156 © 2009 Materials Research Society          1156-D08-06-F06-06

## Power Delivery, Signaling and Cooling for 3D Integrated Systems

Muhannad S. Bakir[1] and Gang Huang[2]

[1] Georgia Institute of Technology, {muhannad.bakir@mirc.gatech.edu}
[2] Was with Georgia Institute of Technology, currently with Intel

### ABSTRACT

Three-dimensional (3D) integration of ICs provides unique opportunities to improve bandwidth, latency, and power dissipation bottlenecks of interconnects (both on- and off-chip). However, while 3D IC integration improves signal interconnection, it also presents new challenges, especially in power delivery and cooling ("thermal interconnects"). The focus of this paper is on some of the key challenges and promising technologies to address power delivery, cooling, and signaling in a 3D stack of logic ICs.

### INTRODUCTION

The information revolution has been the most important economic event of the past century and its most powerful driver has been the silicon integrated circuit (IC). Over the past fifty years, the migration from BJT to CMOS technology combined with transistor scaling has produced exponential benefits in microchip productivity and performance. However, as gigascale silicon technology progresses beyond the 45 nm node, the performance of a system-on-a-chip (SoC) has lagged by progressively greater margins to reach the "intrinsic limits" of each particular generation of technology. A root cause of this "lag" is the fact that the capabilities of monolithic silicon technology per se have vastly surpassed those of the ancillary or supporting technologies that are essential to the full exploitation of a high performance SoC, especially in areas of cooling [1-4], off-chip signaling [2, 3, 5-7], and power delivery [1, 2, 7]. The need for ever greater off-chip bandwidth will be especially problematic as the number of cores on a microprocessor increases [5, 7, 8]. Revolutionary silicon ancillary technologies are needed to address these challenges. Of course, innovation in silicon ancillary technologies will have to be done in parallel with continued innovations at the chip level (improved scaled transistors and interconnects) as well as system architecture among other things [7, 9].

Three-dimensional system integration can be used either to (for example) partition a single chip into multiple strata to reduce on-chip global interconnect length [10] and/or used to stack chips that are homogenous or heterogeneous. An example of 3D stacking of homogenous chips is memory chips, while an example of heterogeneous chip stacking is memory and microprocessor chips. However, while 3D technology can be used to enhance communication between ICs (larger bandwidth, lower latency, and lower energy per bit), it also presents a number of challenges. Aside from issues relating to manufacturing, power delivery [11] and cooling [3] of a stack of logic chips presents many challenges. Simply put, it is difficult enough to cool and deliver power to a single processor today. Stacking multiple processors and memory chips, for example, presents formidable challenges that require advanced silicon ancillary technologies.

The focus of this paper is mainly on power delivery and cooling of 3D stack of logic ICs. The paper begins with a brief survey of technologies to enable micro-liquid cooling of stacked ICs as well as off-stack optical links. Next, the paper discusses challenges and possible solutions for power delivery to a stack of ICs.

## NOVEL SILICON ANCILLARY TECHNOLOGIES

Overview

Figure 1 illustrates a schematic illustration of the 3D system integration methodology under consideration [12-14]. Each silicon die in the 3D stack contains the following features: 1) a monolithically integrated microchannel heat sink; 2) through-silicon electrical (copper) vias (E-TSVs) and through-silicon fluidic (hollow) vias (F-TSVs); 3) solder bumps (electrical I/Os) and microscale polymer pipes (fluidic I/Os) on the side of the chip opposite to the microchannel heat sink. Microscale fluidic interconnection between strata is enabled by the combination of through-wafer fluidic vias and polymer pipe I/O interconnects. The chips are designed such that when they are stacked, each chip makes electrical and fluidic interconnection to the dice above and below. Consequently, power delivery and signaling can be supported by the electrical interconnects (solder bumps and copper TSVs), and heat removal for each stratum can be supported by the fluidic I/Os and microchannel heat sink. In order to support large off-stack bandwidth, optical I/O may be integrated on the bottom most chip in the stack (Figure 2). Optical TSVs [15] may also be integrated to provide unusual flexibility to system integration. In other words, the 3D technology under consideration combines all critical interconnect functions (power, signal, and thermal) for a gigascale system [16].

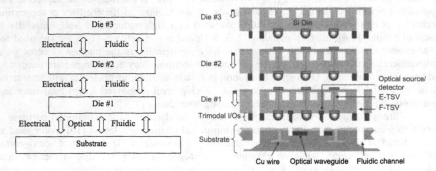

**Figure 1:** Schematic illustration of the 3D system integration technology under consideration.

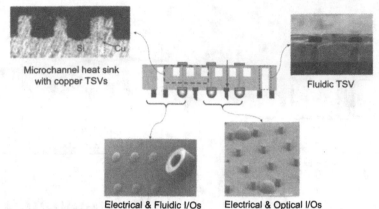

**Figure 2:** Photomicrographs of key interconnect elements of the 3D system.

Optical I/Os

The use of flexible surface-normal optical waveguides, or optical pins, have been proposed as a means of addressing the shortcomings of free-space optical I/O interconnections [17]. The height separation between the chip and the substrate has minimal effect on the optical power received at the photodetector (except for losses through the polymer pin) because the light is tightly confined within the cross-sectional area of the pin. Although we consider using polymeric materials with relatively high optical absorption losses for the fabrication of the optical pins, due to their very short length (height), the optical transmission losses through the pins are small [17, 18]. The optical pins are designed to be mechanically compliant (flexible). The low elastic modulus of the polymer and air cladding of the waveguide contribute to the flexible nature of the optical pins. As a result, the lateral misalignment induced by chip-substrate CTE mismatch is compensated by the mechanical compliance of the optical pins. Thus, optical interconnection and alignment are maintained at all times between the optical components on the chip and substrate due to the mechanical compliance of the optical pins.

Figure 3 illustrates the experimental setup used to characterize the surface-normal optical coupling efficiency of the pins. A fiber was scanned in the X-axis and in the Y-axis across the endface of the pin and across the surface of the aperture (at a Z-axis distance equal to the pin's height). The relative transmitted optical intensity measurements of 50x150 μm optical pins and 50 μm optical apertures are plotted in Figure 3. The transmitted intensities are normalized to the maximum transmission at the center of the aperture without a pin. The X- and Y-axis scans are essentially equal due to the radial symmetry of the light source and the pins. The difference between the coupling efficiency of the two measurements (using data from the X-axis scan) is plotted in Figure 3. The data demonstrate that at the 0 μm displacement position, the optical pins enhance the coupling efficiency by approximately 2 dB when compared to direct coupling into the aperture. At distances of ±25 μm away from the center, the optical coupling improvement due to the pin exceeds 4 dB. The 4 dB coupling improvement is significantly larger than the 0.23 dB excess loss of the pins [17, 18], which demonstrates the benefits of the pins. When 30x150 μm pins (optical aperture 30 μm in diameter) were tested, the coupling efficiency improved by 3 to 4.5 dB [17].

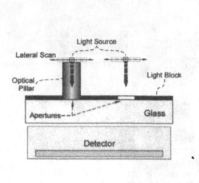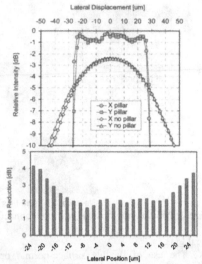

**Figure 3:** The transmitted optical intensity as a function of light source lateral position above the 50x150 μm optical pin (pillar) and aperture.

The polymer pins have also been shown to provide optical displacement compensation. The loss reduction is less than 1 dB up to ~15 μm displacement, while it increases up to 4 dB at 30 μm [17, 18]. This is significant since a limited loss budget is available in typical systems for misalignments/assembly to maintain proper operation. The 4 dB pin-assisted loss reduction at the 30 μm displacement can decrease the BER by possibly few orders of magnitude. Thus, the optical pins provide a method of reducing optical coupling losses caused by thermomechanically induced misalignment between the CTE mismatched chip and substrate.

Fluidic I/Os
    As with the optical I/Os, the fluidic I/Os are flip-chip compatible and are batch-fabricated at the wafer-level [19]. In fact, the optical and fluidic I/Os are batch fabricated simultaneously using the same polymer [16]. This approach is radically different from previously reported research [20, 21] in the area of fluidic interconnects for ICs.
    Figure 3 plots the temperature of the coolant (DI water) at the substrate inlet and outlet and the average chip temperature when 75 W/cm² is applied (for a single die bonded onto a substrate). Under a relatively large flow rate (~104 ml/min), the average temperature rise is 12.7 °C and the corresponding thermal resistance for the chip is approximately 0.28 °C/W [19]. As shown in Figure 3, during testing, the supply power was toggled to verify the consistency of the measurement results. As the microchannel heat sink was not optimized, lower thermal resistance and the cooling of higher power density can be achieved. In fact, the first example of microchannel liquid cooling demonstrated a junction-to-ambient thermal resistance of 0.09 °C/W and the cooling of 790 W/cm² [22], which demonstrates ability to cool hot spots (up to 400 W/cm² in some processors). In this work, we focus on the integration and implementation of the

fluidic interconnect network rather than on the microchannel heat sink. The novel feature of the research is delivering and extracting a liquid coolant from a microchannel heat sink in a way that is compatible with CMOS process technology and conventional chip I/O technology.

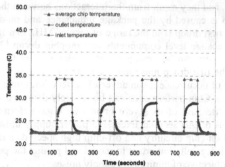

**Figure 4:** Measured average chip temperature at a flow rate of 104 ml/min (for a single die bonded onto a substrate).

In order to achieve high heat transfer, low thermal resistance, and low pressure drop, a relatively tall microchannel heat sink is typically needed (~250 µm, for example). As a result, this necessitates a thick silicon wafer and is different from other 3D integration technologies, which seek to polish the silicon wafer to as small a thickness as possible before wafer handling and mechanical strength become limiters [13, 14]. As a result, this presents interesting set of constraints on the fabrication and integration of E-TSVs. Cross-sectional optical image of fabricated electrical TSVs in a silicon wafer with a microchannel heat sink is shown in Figure 2 [13]. The aspect ratio of the microchannel can be varied to meet the thermal resistance and pressure drop of different applications [13, 22]. TSVs with aspect ratios of 30:1 and greater have been demonstrated [23]. Ultimately, the number of electrical TSVs that can be placed will impact the connectivity between tiers in a stack, including power delivery. In order to address the latter issue, in the next section, we discuss power delivery for 3D ICs.

## POWER DELIVERY FOR 3D SYSTEMS

Power consumption of GSI chips is increasing at an alarming rate [24-26]. The increasingly faster devices packed at unprecedented densities result in high current densities. Although the scaling of the supply voltage has slowed down in recent years, the logic on the integrated circuit (IC) continues to become increasingly sensitive to any supply voltage change because of the decreasing clock cycle and therefore noise margin. With this trend, power supply noise, the voltage fluctuation on power delivery networks, has become a significant factor that can substantially influence the overall system performance. As a result, the design of power delivery systems becomes a very important and challenging task. Therefore, understanding the complicated power delivery networks and supplying clean power to microprocessors is of great significance [25-26].

IR-drop and $\Delta I$ noise are the two main components of the power supply noise. IR-drop results from the supply current passing through the parasitic resistance of the power distribution networks. $\Delta I$ noise is caused by the inductance of the power delivery system and becomes important when a group of circuits switch simultaneously. Power supply noise consists of three distinct voltage droops [27], and they result from the interactions between the chip, package, and board [27]. The first droop is caused by the package inductance and on-die capacitance. The resonance frequency of the first droop is in the range of tens of MHz to a few hundred of MHz depending on the size of package level components and on-chip decap [28]. Because putting additional on-chip decap is very costly, among the three droops, the first droop is the most difficult one to suppress. The first droop noise has the largest magnitude. Even though the first droop has the smallest time of occurrence it can adversely affect GSI circuits as its duration can be tens of nano seconds (ns). Chip performance can be severely degraded when the first droop affects some critical paths. Because of its severe impact on high-performance chips, the first droop is thus the main focus of this section. Excessive power supply noise can lead to severe performance degradation of on-chip circuitry and off-chip high speed data links, and even result in logic failures [25]. Thus it is vitally important to model and predict the performance of power delivery networks with the objective of minimizing supply noise.

On-chip power distribution networks consist of global and local networks. Global power distribution networks carry the supply current and distribute power across the chip. Local networks deliver the supply current from global networks to the active devices. Global networks contribute most of the parasitics, and thus are the main concern of this paper. For global distribution networks, the most common way is to use a grid made of orthogonal interconnects routed on separate metal levels connected through vias [29].

Power delivery and design implications for 3D systems

3D chip stacks have been used in commercial products though today's applications are mainly focused on low power portable devices, such as flash memories and wireless chips. At the high performance end, industry has already started to pave the way for microprocessor stacking and microprocessor-memory stacking, which will extend Moore's Law beyond its expected limits and help break the bottleneck of the memory bandwidth problem for multi-core microprocessors [5]. Through-silicon-vias and micro-bumps are key technologies to fulfill 3D chip stacks for high performance applications. They eliminate the need for long-metal wires that connect today's 2-D chips. These TSVs and micro-bumps enable multiple chips to be stacked together allowing greater amount of information to be passed between them.

However, stacking multiple high-performance dice may result in severe power integrity problems. If multiple high power microprocessors are stacked together and flip-chip technology for 3D chip stacking is used, several hundred amperes of current (or even more) may need to be delivered through limited footprint area. Also, the supply current flows through the micro-bumps and narrow TSVs that may exhibit large parasitic inductance. These may potentially lead to a large $\Delta I$ noise if stacked chips switch simultaneously. Thus, power distribution networks in 3D systems need to be accurately modeled and carefully designed. In [11], analytical models are derived to describe the frequency-dependent characteristics of the power supply noise in each chip in the 3D stack and to obtain physical insight into the rather complex power delivery networks in 3D systems.

The model in [11] is derived from the simplified circuit model of the power distribution network in 3D systems shown in Figure 5. A wire between two nodes on the $i$-th die is simply modeled as a lumped resistance $R_{si}$. The decoupling capacitance per unit area of the $i$-th die is represented by $C_{di}$. The current density for an active block of Die $i$ is represented by $J_i(s)$ in the Laplace domain. Inductance $L_p$ is the per pad loop inductance associated with the package, connected to the bottom-most die (Die 1). Each E-TSV is modeled as connected inductor $L_{via}$ and resistor $R_{via}$ in series (this includes the parasitics of the micro-bumps when they are used between dice). Symbols $\Delta x$ and $\Delta y$ represent the distances between two adjacent power (or ground) nodes in the same wiring level for $x$ and $y$ directions, respectively.

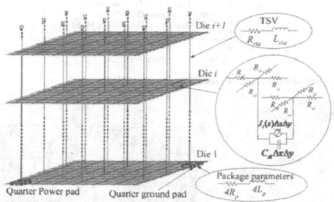

**Figure 5**: Simplified circuit model for 3D stacked system.

In general, if only one die is switching, the noise is smaller than the single chip case (2-D case), because the switching die can use the decaps of those non-switching dice in the 3D stack [11]. However, normally the activities of the two blocks with the same footprints are highly correlated because an important purpose of 3D integration is to put the blocks that communicate most as close to each other as possible. Therefore, we must consider the worst-case scenario when all the functional blocks sharing the same footprint switch simultaneously (Figure 6). If the total number of dice is increased and the noise levels of the topmost and bottommost dice are examined, it can be seen that when all dice are switching the noise produced in a 3D integrated system is unacceptable when compared to a single chip case. This is especially true for the topmost die where the noise level changes dramatically (180 mV for the single die case as opposed to 790 mV for the 10 dice case). Even for the bottommost die, methods of suppressing the noise need to be identified.

175

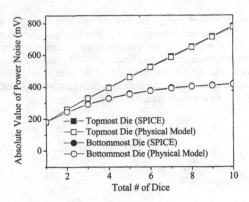

**Figure 6**: All dice switching and increasing total number of dice.

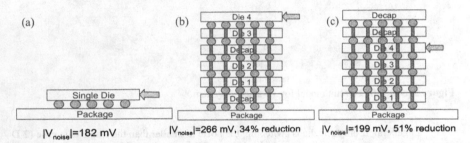

(a)

(b)

(c)

Single Die

$|V_{noise}|$=182 mV      $|V_{noise}|$=266 mV, 34% reduction      $|V_{noise}|$=199 mV, 51% reduction

**Figure 7**: Effect of adding two "decap" dice. (a) Single die switching; (b) One "decap" die at the bottom and the other in the middle; (c) Both "decap" dice on the top.

If we can use a whole die as decap (100% area is occupied by decap) and stack the "decap die" with other dice, the noise can be suppressed to some extent. Figure 7 illustrates the power supply noise when two decap dice are used in the stack (and compared to the single die case). By putting the two decap dice on top, we can suppress the noise to the level of a single chip. Instead of adding a decap die, it will be more efficient if high-k material is used between the power and ground planes (on-chip).

Now, we explore the impact of the number of both P/G pads and TSVs in each die on the power supply noise. When they are both increased, the power supply noise is greatly reduced and even reaches the level of a single chip, as shown in Figure 8. Thus, this demonstrates the importance of large number of TSVs in 3D stack of chips. The inductance of the package is the dominant part throughout the whole power delivery path for the first droop noise. Therefore, the power integrity problem needs an I/O solution that can provide high-density interconnection without sacrificing the mechanical attributes needed for reliability. Of course, electromigration and losses through the power delivery networks need to designed and accounted for as well.

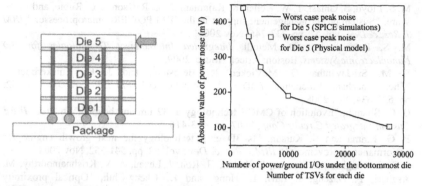

**Figure 8**: Effect of adding more TSVs and power/ground I/Os.

## Conclusion

In order to address the ever increasing adverse effects of conventional silicon ancillary technologies on the performance of CMOS nanosilicon technology, this paper describes the implementation of fully-compatible and wafer-level batch fabricated electrical, optical, and fluidic, or 'trimodal', interconnects. It is proposed that the electrical I/Os be used for power delivery and signaling, the optical I/Os for massive off-stack bandwidth, and the fluidic I/Os (with integrated back-side heat sink) for heat removal. The trimodal I/Os are flip-chip compatible making them compatible with current assembly infrastructure and can be extended to enable the stacking of high-performance (high-power) microprocessors. Liquid cooling for a 3D stack of high-performance chips is also discussed. Challenges in power delivery are also explored for 3D systems. The supply current flowing through the microbumps and narrow through-silicon-vias (TSVs) may have large parasitics. This may potentially lead to a large ΔI noise if stacked chips switch simultaneously. The relationship between the power supply noise, decap insertion, power/ground I/O allocation, and TSVs allocation are discussed. Schemes for reducing the power supply noise in 3D integrated systems are also proposed and their impact on future 3D system designs are also emphasized.

## Acknowledgements

The authors acknowledge the support of the Interconnect Focus Center, one of five research centers funded under the Focus Center Research Program, a DARPA and Semiconductor Research Corporation program. This work is also supported by the National Science Foundation under Grant No. 0701560.

# References

[1]    M. S. Floyd, S. Ghiasi, T. W. Keller, K. Rajamani, F. L. Rawson, J. C. Rubio, and M. S. Ware, "System power management support in the IBM POWER6 microprocessor," *IBM J. Res. & Dev.,* vol. 51, pp. 733-746, Nov 2007.

[2]    M. S. Bakir and J. D. Meindl, *Integrated Interconnect Technologies for 3D Nanoelectronic Systems.* Boston: Artech House, 2009.

[3]    S. M. Sri-Jayantha, G. McVicker, K. Bernstein, and J. U. Knickerbocker, "Thermomechanical modeling of 3D electronic packages," *IBM J. Res. & Dev.,* vol. 52, pp. 623-634, Nov 2008.

[4]    G. G. Shahidi, "Evolution of CMOS technology at 32 nm and beyond," in *Proc. IEEE Custom Integrated Circuits Conf.,* 2007, pp. 413-416.

[5]    P. G. Emma and E. Kursun, "Is 3D chip technology the next growth engine for performance improvement?," *IBM J. Res. & Dev.,* vol. 52, pp. 541-552, Nov 2008.

[6]    J. E. Cunningham, Z. Xuezhe, I. Shubin, H. Ron, J. Lexau, A. V. Krishnamoorthy, M. Asghari, F. Dazeng, J. Luff, L. Hong, and K. Cheng-Chih, "Optical proximity communication in packaged SiPhotonics," in *IEEE Int. Conf. Group IV Photonics,* 2008, pp. 383-385.

[7]    S. Borkar, "Thousand core chips-A technology perspective," in *Proc. ACM/IEEE Design Automation Conference,* 2007, pp. 746-749.

[8]    L. Polka, H. Kalyanam, G. Hu, and S. Krishnamoorthy, "Package technology to address the memory bandwidth challenge for tera-scale computing," *Intel Technol. J.,* vol. 11, pp. 197-205, 2007.

[9]    J. Parkhurst, J. Darringer, and B. Grundmann, "From single core to multi-core: preparing for a new exponential," in *IEEE/ACM Int. Conf. Computer-Aided Design,* 2006, pp. 67-72.

[10]   J. W. Joyner, P. Zarkesh-Ha, and J. D. Meindl, "Global interconnect design in a three-dimensional system-on-a-chip," *IEEE Tran. Very Large Scale Integration (VLSI) Systems,* vol. 12, pp. 367-372, 2004.

[11]   G. Huang, M. Bakir, A. Naeemi, H. Chen, and J. D. Meindl, "Power delivery for 3D chip stacks: physical modeling and design implication," in *Proc. IEEE Conf. on Electrical Performance of Electronic Packaging,* 2007, pp. 205-208.

[12]   C. K. King, D. Sekar, M. S. Bakir, B. Dang, J. Pikarsky, and J. D. Meindl, "3D Stacking of Chips with Electrical and Microfluidic I/O Interconnects," in *Proc. Electronics Components and Technol. Conf.,* 2008.

[13]   D. Sekar, C. King, B. Dang, T. Spencer, H. Thacker, P. Joseph, M. S. Bakir, and J. D. Meindl, "A 3D-IC Technology with Integrated Microchannel Cooling," in *Proc. Int. Interconnect Technol. Conf.,* 2008, pp. 13-15.

[14]   M. S. Bakir, C. King, D. Sekar, H. Thacker, B. Dang, G. Huang, A. Naeemi, and J. D. Meindl, "3D heterogeneous integrated systems: liquid cooling, power delivery, and implementation," in *Proc. IEEE Custom Integrated Circuits Conf.,* 2008.

[15]   H. Thacker, O. Ogunsola, A. Carson, M. Bakir, and J. Meindl, "Optical through-wafer interconnects for 3D hyper-integration," in *Proc. IEEE Lasers & Electro-Optics Society Annual Meeting,* 2006, pp. 28-29.

[16]    M. S. Bakir, B. Dang, and J. D. Meindl, "Revolutionary nanosilicon ancillary technologies for ultimate-performance gigascale systems," in *Proc. IEEE Custom Integrated Circuits Conf.*, 2007, pp. 421-428.

[17]    M. S. Bakir, A. L. Glebov, M. G. Lee, P. A. Kohl, and J. D. Meindl, "Mechanically Flexible Chip-to-Substrate Optical Interconnections Using Optical Pillars," *IEEE Trans. on Advanced Packaging*, vol. 31, pp. 143-153, 2008.

[18]    A. L. Glebov, D. Bhusari, P. Kohl, M. S. Bakir, J. D. Meindl, and M. G. Lee, "Flexible pillars for displacement compensation in optical chip assembly," *IEEE Photonics Technol. Lett.*, vol. 18, pp. 974-976, 2006.

[19]    B. Dang, "Integrated Input/Output Interconnection and Packaging for GSI," Ph. D. Thesis, Georgia Institute of Technology, 2006.

[20]    H. Y. Zhang, D. Pinjala, T. N. Wong, and Y. K. Joshi, "Development of liquid cooling techniques for flip chip ball grid array packages with High Heat flux dissipations," *IEEE Trans. Components and Packaging Technol.*, vol. 28, pp. 127-135, 2005.

[21]    E. G. Colgan, B. Furman, A. Gaynes, W. Graham, N. LaBianca, J. H. Magerlein, R. J. Polastre, M. B. Rothwell, R. J. Bezama, R. Choudhary, K. Marston, H. Toy, J. Wakil, and J. Zitz, "A practical implementation of silicon microchannel coolers for high power chips," in *Proc. IEEE Semiconductor Thermal Measurement and Management Symp.*, 2005, pp. 1-7.

[22]    D. B. Tuckerman and R. F. W. Pease, "High-performance heat sinking for VLSI," *IEEE Electron Dev. Lett.*, vol. 2, pp. 126-129, 1981.

[23]    J. H. Wu, J. Scholvin, and J. A. del Alamo, "A through-wafer interconnect in silicon for RFICs," *IEEE Trans. Electron Devices*, vol. 51, pp. 1765-1771, 2004.

[24]    J. D. Meindl, "Low power microelectronics: retrospect and prospect," *Proceedings of IEEE*, vol. 83, pp. 619-635, Apr. 1995.

[25]    M. Swaminathan and E. Engin, *Power Integrity: Modeling and Design for Semiconductor and Systems*, Prentice Hall, 1st Edition, 2007.

[26]    H. Zheng, B. Krauter and L.T. Pileggi, "Electrical modeling of integrated-package power/ground distributions," *IEEE Design and Test of Computer*, vol. 20, no. 3, pp. 23-31, May-June, 2003.

[27]    K. L. Wong, T. Rahal-Arabi, M. Ma, and G. Taylor, "Enhancing microprocessor immunity to power supply noise with clock-data compensation," *IEEE Journal of Solid-State Circuits*, vol. 41, no. 4, April, 2006.

[28]    W. D. Becker, J. Eckhardt, R. W. Frech, G. A. Katopis, E. Klink, M. F. McAllister, T. G. MacNamara, P. Muench, S. R. Richter, and H. H. Smith, "Modeling, simulation, and measurement of mid-frequency simultaneous switching noise in computer systems," *IEEE Transactions on Components, Packaging, and Manufacturing Technology*, part B, vol. 21, pp. 157–163, May 1998.

[29]    A. Dharchoudhury, R. Panda, D. Blaauw, R. Vaidyanathan, "Design and analysis of power distribution networks in PowerPC microprocessors," in *Proc. IEEE Design Automation Conf.*, 15-19 June 1998 pp: 738 – 743.

Mater. Res. Soc. Symp. Proc. Vol. 1156 © 2009 Materials Research Society                1156-D08-07-F06-07

# Copper Deposition Technology for Thru Silicon Via Formation Using Supercritical Carbon Dioxide Fluids in a Flow-Type Reaction System

Masahiro Matsubara and Eiichi Kondoh

Graduate School of Engineering, University of Yamanashi,

4-3-11 Takeda, Kofu-city, Yamanashi 400-8511, Japan

## ABSTRACT

A supercritical fluid is a high-pressure medium that possesses both high diffusivity and solvent capabilities. Metal thin films can be deposited in supercritical fluids from an organometallic compound (precursor) through thermochemical reactions. In the present study, we used a technique, aimed at applying to the fabrication of through-silicon vias (TSVs), where copper thin films were deposited in silicon microholes 10 μm in diameter and 350 μm in depth. The temperature and pressure were varied from 180°C to 280°C and 1 MPa to 20 MPa, respectively. The maximum coating depth decreased with deposition temperature, whereas a peak maximum of the depth was observed at around 10 MPa. The temperature and pressure dependences on the coating depth were numerically studied. On the basis of the analysis, a deposition program was modified as to elongate the coating depth.

Keywords: through-silicon via, supercritical carbon dioxide, copper thin film, diffusion constant

## INTRODUCTION

Three-dimensional (3D) die stacking technology offers many potential advantages compared to traditional packaging technologies. For example, the distance between circuits can be significantly reduced, and use of through silicon via (TSV) can increase chip interconnection density compared to that of traditional wire-bonding interconnection. Wafer processing can be simplified to having differently functioned chips such as processors or memory, instead of adding process steps for heterogeneous function, in a conventional two-dimensional (2D) chip. This new 3D silicon technology is emerging at a time when Moore's Law for semiconductor chip scaling seems to be slowing or reaching an end [1].

3D IC applications may require only millions of TSV, in the near future, so that TSVs should have a very high aspect ratio depending on silicon thinning technology. In the TSV fabrication process, vias etched with deep silicon etching are filled with copper. Existing deposition technologies, such as sputtering and electroplating, have limitations in forming a uniform liner and filling TSVs with conductor metal.

We have been investigating Cu thin film deposition technology using supercritical carbon dioxide (scCO$_2$) in order to fill/coat a very high aspect ratio of vias. In this technology vehicle, wherein an organometal complex is dissolved in scCO$_2$, it is reduced to a thin film. Deposition of variety of noble and near-noble metals has been reported from Cu to Pd, Au, Ni, Pt, Rh, and to alloys.[2-6] ScCO$_2$ possesses not only its density and solvent capability as high as liquid, but also has high diffusivity and low viscosity as a gas and has zero-surface tension, which allows scCO$_2$ to penetrated into very small features. These properties make scCO$_2$ an ideal medium for thin film chemical deposition for coating or filling TSV with metal with good coverage/filling capability.

From these viewpoints, we studied the deposition characteristics of copper thin films on the sidewalls of TSV test structures.

## EXPERIMENTAL

Deposition was carried out with a flow-type $scCO_2$ processor that was developed in our laboratory[7,8]. The copper precursor used was copper(II)bis-diisobutylmethanate, abbr. $Cu(dibm)_2$, and was dissolved in an auxiliary solvent (acetone) and injected into $scCO_2$ with a high pressure pump. The standard deposition condition was fixed at a $CO_2$ flowrate of 3.5ml/min ($7.75 \times 10^{-2}$ mol/min), a pressure of 10MPa, a temperature of 220 °C, and a deposition time of 60 min. The temperature was varied from 180 °C to 280 °C and the pressure was varied from 1 to 20 MPa when studying their dependences. The molar ratios of $H_2$, $Cu(dibm)_2$, and acetone to $CO_2$ were 1.53 %, 0.0292% , and 5.29%, respectively. The TSV test structure used had microholes etched in silicon with a diameter of 10 um and depth of 350 μm.

## RESULTS AND DISCUSSION

### Basic film properties

Figure 1 shows a cross-sectional SEM image of Si holes (a) and an EDX spectrum obtained at an entrance of a via. Deposition was carried out at 10 MPa and 240°C. The coating depth, with is defined as the depth Cu film reached, was 90 μm in this particular case. The EDX spectrum exhibited a strong Cu peak along with weak Si, C, and O peaks. The latter two elements originated from the molding resin. The resistivity of this film measured at blanket part was about 2.0 μΩ·cm. The elastic modulus was also measured by nanoindentation, and its value was approximately 130 MPa, close to the bulk value. These show that that the deposited Cu film has high purity.

### Experimental and analytical study of coating depth

The coating depth was measured on optical micrographs of via cross-sections, and the temperature and pressure dependences are shown in Figure 2(a) and 2(b), respectively. It was seen that the coating depth decreased with temperature. Also, the coating depth increased with

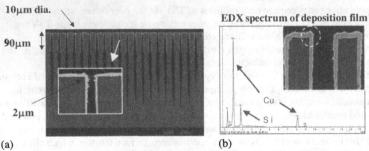

EDX spectrum of deposition film

10μm dia.

90μm

2μm

(a)                                                          (b)

Figure 1      Cross-sectional SEM image of Si holes (a) and an EDX spectrum obtained at a corner of a via.

pressure, peaked at 13 MPa, and then decreased.

Now we will discuss these dependences by using an analytical model. A Péclet number of the scCO$_2$ medium in the Si holes in the order of $10^0$, taking the diameter of hole as the characteristics length of the system and the fluid velocity in the reactor as the characteristic velocity. In this particular case, the diffusive inflow of chemical species and its consumption of the precursor inside via was balanced. This mass balance is represented by using a Thiele model [9],

$$R\rho_{\text{fluid}}D\frac{d^2C_i}{dx^2} - 2\rho_{\text{film}}G(C_i) = 0, \tag{1}$$

where $R$ is the via diameter, $C_i$ is the concentration of species i, $G$ is the growth rate as a function of species i, and the others have their usual meanings. For $G$, we employed the following rate equation:

$$G(C_{\text{pre}}, C_{H_2}) \propto \exp\left(-\frac{3990}{T}\right)\frac{\left(63.4\exp\left(\frac{1850}{T}\right)C_{\text{pre}}\right)\sqrt{4040\exp\left(-\frac{6250}{T}\right)C_{H_2}}}{\left(1+63.4\exp\left(\frac{1850}{T}\right)C_{\text{pre}}+\sqrt{4040\exp\left(-\frac{6250}{T}\right)C_{H_2}}\right)^2} \quad \text{(nm/min)}. \tag{2}$$

This equation was obtained by a non-linear fitting our 43 experiments under the same experimental chemistry [8]. The diffusion constant $D$ was estimated using a Chapman-Enskog equation [10]. The value of $C_i$ was fixed at $x = 0$ (entrance) and $dC/dt = 0$ was set at $x = 350$ μm (bottom), which were used as boundary conditions. Equation 1 was solved numerically with these boundary conditions and compared with the results of the present experiment. The numerical coating depths were defined as the point where $G = 0.17$ nm/min that is roughly equivalent to the skin depth of Cu at (10 nm at $\lambda = 600$ nm) for 60 min deposition.

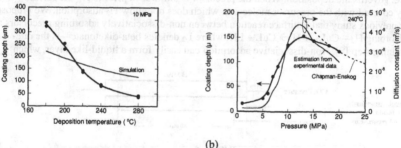

(a)                                                          (b)

Figure 2      Temperature (a) and pressure (b) dependences of coating depth.

Calculations showed agreement with the tendency of the temperature dependence where the coating depth decreased with temperature. This is generally because the growth rate was more temperature dependent than the diffusion constant, and therefore the precursor molecules (H$_2$ as well) could travel longer. In addition, when the temperature was low, $G$, eq. 2, became less concentration dependent, i.e., the reaction chemistry nears 0-order, leading to uniform thickness distribution along the depth.

Our simulation showed a peak in the coating depth at around 10 MPa. The reason for the peaking was that the $G$ increased almost linearly, but $D$ decreased inversely with pressure. However, in the simulation, the pressure dependency was much weaker than that in the experimental results shown as a thick line in Fig. 2(b). We then calculated $D$ back from the experimentally obtained coating depth. The calculated $D$ is displayed as a thin solid line in Fig. 2(b) along with that given by a Chapman-Enskog equation (broken line). Above 10 MPa, $D$ is 3–$4 \times 10^{-8}$ m$^2$/s and is in good agreement with the theory. However, below 10 MPa, $D$ drops off with decreasing pressure and reached to an order of $10^{-9}$ m$^2$/s that is almost equivalent to the value of a liquid. Presumably, apart from the physicochemical mechanism, a kind of stagnation layer was built when the pressure was low, which leads to suppressing diffusive transport.

Coating depth enhancement by dynamic pressurizing
As stated in the end of the previous section, the formation of a stagnation layer can limit the coating depth. The stagnation layer may be built at a higher pressure, therefore it is important to enhance advective transport around vias. For this purpose, developed a new recipe.

In our standard recipe, $H_2$-dissolving $CO_2$ was flown to the reactor and then the precuro was supplied continuously (Fig. 3 above). In this case the coating depth was 143 µm. In a new "breathing" recipe, the pressure was oscillated between 15 MPa and 9 MPa, in order to introduce forced convection. $H_2$ and the precursor were supplied only during the 15 MPa cycle, and the $H_2$ was supplied earlier than and after the precursor supply ($H_2$ first). The precursor was supplied for 10 min, and this cycle was repeated 6 times (Fig. 3 below). The coating depth was expanded to 184 µm (increase by 30 %) as expected.

Interestingly, when the order of $H_2$ and precursor supply was reversed (precursor first), the coating depth was decreased to 58 µm. This shows that 1) the cyclic process is "effective" in terms of changing the deposition environment and that 2) the initial species being introduced in the environment is crucial. The latter implies that the initial species, which adsorbs on the surface, governs the deposition. When the precursor is introduced prior to $H_2$, the surface will be mostly covered with the precursor molecules, which does not allow $H_2$ adsorption. We proposed in our previous study that a surface reaction between non-dissociatively adsorbing precursors and chemisorbing $H_2$—$CuL_2 + H \rightarrow CuL + LH$, where $L_2$ denotes beta-diketonate—is the rate-determining step [8]. Non-dissociative adsorption can easily form a liquid-like layer, which

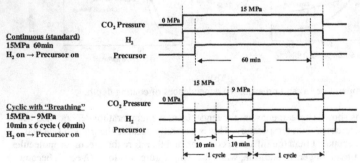

Figure 3        Time charts of fluid supply sequence. Standard continuous (above) and modified "breathing" (below).

seems to correspond to the stagnation layer we discussed above. Therefore, in this deposition technology, promoting preferential adsorption of $H_2$ is important.

## CONCLUSIONS

Copper thin films were deposited in silicon microholes of 10 µm in diameter and 350 µm in depth, by using a technique applying to fabrication through-silicon vias (TSVs) for three-dimensional silicon integrated circuits. The temperature and pressure were varied from 180°C to 280°C and 1 MPa to 20 MPa, respectively. The maximum coating depth decreased with deposition temperature from 350 µm to 40 µm, whereas a peak maximum of the depth was observed at around 10 MPa. The temperature and pressure dependences on the coating depth were discussed using a Thiele model that described the balance between diffusive transport and consumption of the precursor, and fairly strong agreement with the experimental results was found. This shows that the diffusion constants in the silicon microholes were accurately estimated from the observed maximum coating depths, and the presence of a stagnation layer that retarded the transport of species was suggested. In order to suppress the formation of the stagnation layer, a cyclic "breathing" process sequence was developed and was found to be effective.

## ACKNOWLEDGEMENT

We acknowledge that TSV test structures were provided by OMRON Corp., Kyoto, Japan.

## REFERENCES

1. J. U. Knickerbocker, P. S. Andry, B. Dang, R. R. Horton, M. J. Interrante, C. S. Patel, R. J. Polastre, K. Sakuma, R. Sirdeshmukh, E. J. Sprogis, S. M. Sri-Jayantha, A. M. Stephens, A. W. Topol, C. K. Tsang, B. C. Webb, S. L. Wright; *IBM J. Res. Development*, **62**, 556 (2008).
2. X. R. Ye, Y. Lin and C. M. Wai, Chem. Commun., **2003**, 642 (2003).
3. H. Wakayama, N. Setoyama, Y. Fukushima, Adv. Mater., **2003**, 743 (2003).
4. E. Kondoh, Jpn. J. Appl. Phys., **43**, 3928 (2004)
5. A. O'Neil and J. J. Watkins, MRS Bull., **30**, 967 (2005).
6. B. Zhao, T. Momose, and Y. Shimogaki, Jpn. J. Appl. Phys., **45** (2006) L1296.
7. E. Kondoh and J. Fukuda, J. Supercritical Fluids, **44** (2008) 466.
8. M. Matsubara, M. Hirose, K. Tamai, Y. Shimogaki, and E. Kondoh, J. Electrochem. Soc., **156**, H443 (2009)
9. O. Levenspiel, Chemical Reaction Engineering, second ed., p. 472 (John Wiley & Sons, 1972).
10. S. Chapman and T. G. Cowling, The Mathematical Theory of Non-Uniform Gases, third ed., p. 167 (Cambridge UP, London, 1970).

Agarwal, Rahul, 149
Aki, S., 99
Aksenov, German, 39
Anthis, J., 73
Arai, H., 39
Armini, Silvia, 85

Bakir, Muhannad, 169
Baklanov, Mikhail R., 17, 23, 39, 53
Besser, Paul R., 113
Beyer, Gerald, 23
Blonkowski, Serge, 3
Boag, N., 73
Braginsky, O.V., 17
Busawon, Abheesh N., 53

Calbo, Giovanni, 23
Chabal, Yves J., 73
Chen, Fen, 133
Choi, Samuel, 11
Ciofi, Ivan, 23

DeRoest, David, 39
Dixit, Vijay K., 99, 105
Dultsev, F.N., 39
Dziobkowski, Chet, 11

Eslava, Salvador, 53

Fang, Nicholas, 141
Ferreira, Placid, 141
Fujimoto, Koji, 163

Gall, Martin, 121
Gambino, Jeff, 133
Gudmundsson, Jon T., 59
Guedj, Cyril, 3

Hauschildt, Meike, 121
He, John, 133
Hernandez, Richard, 121
Hosaka, S., 99
Hsieh, Tsung-Eong, 65
Hsu, Keng, 141

Huang, Gang, 169

Iacopi, Francesca, 53
Ingason, Arni S., 59
Ito, Kazuhiro, 93
Itoh, H., 99

Jacobs, Kyle, 141
Jordan-Sweet, Jean L., 113

Kanjolia, R., 73
Kesters, Els, 45
Kirschhock, Christine E., 53
Kitada, Hideki, 163
Kohama, Kazuyuki, 93
Koike, Junichi, 99, 105
Kondoh, Eiichi, 181
Kovalev, A.S., 17
Kumar, Anil, 141

Le, Quoc Toan, 45
Lee, Tom C., 133
Lopaev, D.V., 17
Lux, Marcel, 45

Maeda, Nobuyuki, 163
Maekawa, Kazuyoshi, 93
Maestre Caro, Arantxa, 85
Maex, Karen, 53
Magnus, Fridrik, 59
Majeed, Bivragh, 155
Malykhin, E.M., 17
Mankelevich, Y.A., 17
Marsik, Premysl, 39, 45
Martens, Johan A., 53
Matsubara, Masahiro, 181
Matsumoto, K., 99
Mongeon, Steve, 133
Mori, Kenichi, 93
Mosher, Dave, 133
Murakami, Masanori, 93
Murray, Conal E., 113

Nakamura, Tomoji, 163
Neishi, Koji, 99, 105

Odedra, R., 73
Ohba, Takayuki, 163
Olafsson, Sveinn, 59

Pan, Hung-Chun, 65
Park, Sun Kyung, 73
Pawlikowski, Gregory, 31
Pokrinchak, Phil, 133
Prager, L., 45
Proshina, O.V., 17

Radhakrishnan, Bala, 79
Rakhimov, A.T., 17
Rakhimova, T.V., 17
Roodenko, K., 73
Roy, David, 3
Ruythooren, Wouter, 149

Sarma, Gorti, 79
Sato, H., 99
Shamiryan, Denis, 39
Shinosky, Mike, 133
Shirai, Yasuharu, 93

Soussan, Philippe, 155
Suzuki, Kousuke, 163
Sylvestre, Alain, 3

Tai, Leo, 11
Takamure, N., 39
Tezcan, Deniz S., 155
Tökei, Zsolt, 23

Urrutia, Jone, 53

Van Cauwenberghe, Marc, 155
Vasilieva, A.N., 17
Verdonck, Patrick, 39
Vereecke, Guy, 45
Verriere, Virginie, 3
Voloshin, D.G., 17

Wielunski, L., 73
Witt, Christian, 113

Zyryanov, S.M., 17

absorption, 39
additives, 31
adhesion, 99
adhesive, 163
Ag, 141
amorphous, 65, 105
atomic layer deposition, 73

barrier layer, 65, 93, 99
bonding, 149

chemical vapor deposition (CVD)
    (deposition), 39, 99, 105
Cu, 11, 85, 105, 113, 121, 133,
    149, 163, 181

defects, 3
dielectric, 3, 39, 53
    properties, 17, 23, 31, 133
diffusion, 17, 65

electrical properties, 3, 59
electrodeposition, 85, 163
electromigration, 121
electronic material, 59

H, 23

infrared (IR) spectroscopy, 45

kinetics, 121

lithography (removal), 45

metalorganic deposition, 181

microelectro-mechanical (MEMS),
    141, 169
microelectronics, 133, 169
microstructure, 79

nano-indentation, 141

optical properties, 73

packaging, 149, 155, 169
photochemical, 45
plating, 155
polymer, 31
porosity, 17

reactive ion etching, 11, 155
Ru, 73
Rutherford backscattering
    (RBS), 93

self-assembly, 85
simulation, 79
sol-gel, 53

texture, 79
thermal stresses, 113
thin film, 53, 93, 181

water, 23

x-ray
    diffraction (XRD), 59, 113
    photoelectron spectroscopy
        (XPS), 11

Printed in the United States
By Bookmasters

Printed in the United States
By Bookmasters